CONTENTS

3

APPENDICES

INDEXES

Co-operating Organisations

The Institution of Electrical Engineers acknowledges the contribution made by the following organisations in the preparation of this guide.

Association of Manufacturers of Domestic Electrical Appliances
S A MacConnacher BSc CEng MIEE

British Cables Association
C K Reed I Eng MIIE

British Electrotechnical & Allied Manufacturers Association Ltd
R Lewington

British Electrotechnical Approvals Board
P D Stokes MA CEng MRAeS

British Standards Institution
W E Fancourt

City & Guilds of London Institute
H R Lovegrove IEng FIIE

Electrical Contractors' Association
D Locke IEng MIIE ACIBSE

Electrical Contractors' Association of Scotland t/a SELECT
D Millar

Electrical Installation Equipment Manufacturers' Association Ltd
Eur Ing M H Mullins BA CEng MIEE FIIE

Electricity Association Limited
D J Start BSc CEng MIEE

ERA Technology Ltd
M W Coates B Eng

Federation of the Electronics Industry
F W Pearson CEng MIIE

The GAMBICA Association Ltd
K A Morriss BSc CEng MIEE MInstMC

Health & Safety Executive
Eur Ing J A McLean FIEE FIOSH

Institution of Incorporated Engineers
P Tootill IEng MIIE

Lighting Association
K R Kearney IEng MIIE

National Inspection Council for Electrical Installation Contracting

Safety Assessment Federation Limited
J Gorman BSc (Hons) CEng MIEE

Society of Electrical and Mechanical Engineers serving Local Government
C Tanswell CEng MIEE MCIBSE

PREFACE

The On-Site Guide is one of a number of publications prepared by the Institution of Electrical Engineers to simplify some aspects of BS 7671 : 2001 incorporating amendment No 1, 2002 Requirements for Electrical Installations (IEE Wiring Regulations Sixteenth Edition). BS 7671 is a joint publication of the British Standards Institution and the Institution of Electrical Engineers.

The scope generally follows that of BS 7671. It includes material not included in BS 7671, provides background to the intentions of BS 7671, and gives other sources of information.

However, this guide does not ensure compliance with BS 7671. It is a simple guide to the requirements of BS 7671, and electricians should always consult BS 7671 to satisfy themselves of compliance.

It is expected that persons carrying out work in accordance with this guide will be competent to do so.

Electrical installations in the United Kingdom which comply with the IEE Wiring Regulations, BS 7671, should also comply with the Statutory Regulations such as the Electricity at Work Regulations 110-01-01 1989. It cannot be guaranteed that BS 7671 complies with all relevant Regulations and it is stressed that it is essential to establish what statutory and other Regulations apply and to install accordingly. For example, an installation in Licensed Premises may App 2(vi) have requirements different from or additional to BS 7671 and these will take precedence over BS 7671.

FOREWORD

This Guide is concerned with limited application of BS 7671 in accordance with para 1.1 Scope. Part 1

BS 7671 and the On-Site Guide are not design guides. It is essential to prepare a schedule of the work to be done prior to commencement or alteration of an electrical installation and to provide all necessary information and operational manuals of any equipment supplied to the user on completion.

Any specification should set out the detailed design and provide sufficient information to enable competent persons to carry out the installation and to commission it. The specification must include a description of how the system is to operate and all of the design and operational parameters.

The specification must provide for all the commissioning procedures that will be required and for the production of any operational manual.

It must be noted that it is a matter of contract as to which person or organisation is responsible for the production of the parts of the design, specification and any operational manual.

The persons or organisations who may be concerned in the preparation of the specification include:

The Designer(s)
The Installer(s)
The Supplier of Electricity
The Installation Owner and/or User
The Architect
The Fire Prevention Officer
The Planning Supervisor
All Regulatory Authorities
Any Licensing Authority
The Health and Safety Executive

In producing the specification advice should be sought from the installation owner and/or user as to the intended use. Often, such as in a speculative building, the detailed intended use is unknown. In those circumstances the specification and/or the operational manual must set out the basis of use for which the installation is suitable.

Precise details of each item of equipment should be obtained from the manufacturer and/or supplier and compliance with appropriate standards confirmed.

The operational manual must include a description of how the system as installed is to operate and all commissioning records. The manual should also include manufacturers' technical data for all items of switchgear, luminaires, accessories, etc and any special instructions that may be needed. The Health and Safety at Work etc Act 1974 Section 6 and the Construction (Design and Management) Regulations 1994 are concerned with the provision of information. Guidance on the preparation of technical manuals is given in BS 4884 (Specification for technical manuals) and BS 4949 (Recommendations for the presentation of technical information about products and services in the construction industry). The size and complexity of the installation will dictate the nature and extent of the manual.

ON-SITE GUIDE

SECTION 1.   INTRODUCTION

1.1   Scope

This Guide is for electricians (skilled persons). It covers the following installations:

(a)   domestic installations generally, including off-peak supplies, and supplies to associated garages, out-buildings and the like

(b)   industrial and commercial single- and three-phase installations where the distribution board(s) or consumer unit is located at or near the supplier's cut-out.

Note: Special Installations or Locations (Part 6 of BS 7671) are generally excluded from this Guide. Advice is given on installations in locations containing a bath or shower (8.1), temporary and garden buildings etc (8.3) and personal computer circuits in office locations (8.4).

Part 6

This Guide is restricted to installations :

313-01-01

(i)   at a supply frequency of 50 Hertz

(ii)   at a nominal single-phase voltage of 230 V a.c. single-phase and 230/400 V a.c. three-phase

(iii)   fed through a supplier's cut-out having a fuse or fuses to BS 1361 Type II or through fuses to BS 88-2 or BS 88-6

(iv)   with a maximum value of the earth fault loop impedance outside the consumer's installation as follows:

Earth return via sheath (TN-S system): 0.8 ohm

Earth return via combined neutral and earth conductor (TN-C-S system): 0.35 ohm

TT systems: 21 ohms excluding consumer's earth electrode

This Guide contains information which may be required in general installation work, e.g. conduit and trunking capacities, bending radii of cables.

This Guide introduces the use of conventional circuits, which are discussed in Section 7.

Because of simplification this Guide may not give the most economical result.

This Guide is not a replacement for BS 7671, which should always be consulted. Defined terms according to Part 2 of BS 7671 are used in this Guide.

In conformance with the definitions of BS 7671, throughout this Guide the term 'live part' is used to refer to a conductor or conductive part intended to be energised in normal use, including a neutral conductor. For convenience in use, and in accordance with current UK manufacturing practice, the Part 2 terminals of electrical equipment shown in Figs 10.1 to 10.5 figs 3 to 7 have been identified by the letters L, N and E.

Further information is available in the series of Guidance Notes published by the Institution.

For new domestic installations and major refurbishments account should be taken of the recommendations in Approved Document B, issued as guidance on the Building Regulations 1991. Part B1, Section 1 advises that, if dwellings are not protected by an automatic fire detection and alarm system to Part 6 of BS 5839, a suitable number of mains operated self-contained smoke alarms to BS 5446 be installed.

In Scotland the Building Standards (Scotland) Regulations 1990 apply and installations must comply with The Scottish Office Technical Standards.

Before starting work on an installation that requires a new supply, the electrician should obtain the following information from the supplier:

(i)	the number of phases to be provided	312-02-01
(ii)	the supplier's requirement for cross-sectional area and length of meter tails	313-01-01(iii)
(iii)	the maximum prospective fault current (pfc) at the supply terminals	313-01-01(iii)
(iv)	the maximum earth loop impedance (Z_e) of the earth fault path outside the consumer's installation	313-01-01(iv)
(v)	the type and rating of the supplier's fusible cut-out or protective device	313-01-01(vi)
(vi)	the supplier's requirement regarding the size of main equipotential bonding	547-02-01
(vii)	the earthing arrangement and type of system	312-03-01
(viii)	the arrangements for the incoming cable and metering.	313-01-01(v)

For existing installations, electricians should satisfy themselves as to the suitability of the supply including the earthing arrangement.

SECTION 2.   THE SERVICE POSITION

2.1   General Layout of Equipment

The general layout of the equipment at the service position is shown in Figs 2a and 2b.

2.2   Function of Components:

(i)   Distributor's Cut-out

This will be sealed to prevent the fuse being withdrawn 313-01-01(vi) by unauthorised persons. When the meter tails and consumer unit are installed in accordance with the requirements of the distributor the cut-out may be assumed to provide fault current protection up to the consumer's main switch.

(ii)   Supplier's Meter

This will be sealed by the supplier to prevent interference by unauthorised persons.

(iii)   Meter Tails

These are part of the consumer's installation. They 473-02-04(iv) should be insulated and sheathed or insulated and 521-07-03 enclosed in conduit or trunking.

Polarity should be indicated by the colour of the 514-06 insulation and the minimum cable size should be 25 mm^2. The supplier may specify the maximum length and the minimum cross-sectional area (see 1.2(ii)).

Where the meter tails are protected against fault current 473-02-04(iv) by the supplier's cut-out the method of installation, maximum length and minimum cross-sectional area must comply with the requirements of the supplier.

Fig 2a: Layout when the supplier does not provide a main switch

Note: Earthing arrangements have been omitted for clarity. Tails between the meter and the consumer's installation are provided by the consumer

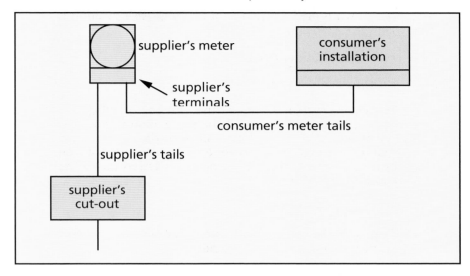

Fig 2b: Layout when the supplier does provide a main switch

Note: Earthing arrangements have been omitted for clarity. Tails between any main switch provided by the supplier and the consumer's installation are provided by the consumer

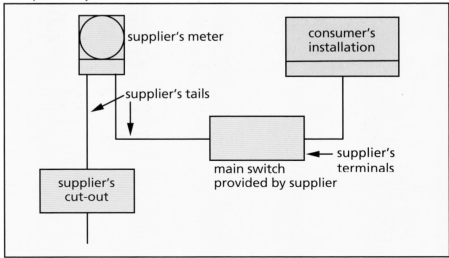

(iv) *Supplier's Switch*

Some suppliers may provide and install a suitable switch between the meter and the consumer unit. This permits 476-01-01 the supply to the installation to be interrupted without withdrawing the supplier's fuse in the cut-out.

(v) *Consumer's Controlgear*

A consumer unit is for use on single-phase installations. Part 2 It contains a double-pole main switch and fuses or circuit-breakers (cbs) and perhaps residual current devices (RCDs) or residual current breakers with integral overcurrent protection (RCBOs) for the protection of each final circuit. Alternatively, a separate main switch and distribution board may be provided.

SECTION 3.   PROTECTION

3.1   Types of Protective Device(s)

The consumer unit (or distribution board) contains devices for Ch 43
the protection of the final circuits against:

(i)	overload	433
(ii)	short-circuit	434
(iii)	earth fault.	434

Functions (i) and (ii) are carried out usually by one device, a fuse
or circuit-breaker.

Function (iii) may be carried out by the fuse or circuit-breaker 413-02-04
provided for functions (i) and (ii), or by an RCD.

An RCBO being a combined circuit-breaker and RCD will carry
out functions (i), (ii) and (iii).

3.2   Overload Protection

Overload protection is given by the following devices:

Fuses to BS 88-2.1 or BS 88-6; BS 1361 and BS 3036; App 3
miniature circuit-breakers to BS 3871-1 Types 1, 2 and 3;
circuit-breakers to BS EN 60898 types B, C and D; and
residual current circuit-breakers with integral
overcurrent protection (RCBOs) to BS EN 61009-1.

3.3   Fault Current Protection

When a consumer unit to BS EN 60439-3 or BS 5486 : Part 13, 473-02-04
or a fuseboard having fuselinks to BS 88-2.1 or BS 88-6 or BS 1361
is used, then fault current protection will be given by the
overload protective device.

For other protective devices the breaking capacity must be
adequate for the prospective fault current at that point.

3.4   Protection Against Electric Shock

(i)   Direct Contact   412

Electrical insulation and enclosures and barriers give 412-01-01
protection against direct contact. Non-sheathed
insulated conductors must be protected by conduit or 521-07-03

trunking or be within a suitable enclosure. A 30 mA RCD may be provided to give supplementary protection against direct contact, but must not be relied upon for primary protection.

412-06

(ii)   *Indirect Contact*

413

Protection against indirect contact is given by limiting to safe values the magnitude and duration of voltages that may appear under earth fault conditions between simultaneously accessible exposed-conductive-parts of equipment, and between them and extraneous-conductive-parts or earth. This may be effected by the:

(a)   co-ordination of protective devices and circuit impedances, or

413-02-04

(b)   use of RCDs to limit the disconnection time, or

413-02-07

(c)   use of Class II equipment or equivalent insulation.

413-03

(iii)   *SELV and PELV*

SELV
Separated extra-low voltage (SELV) systems

411-02

(a)   are supplied from isolated safety sources such as a safety isolating transformer to BS 3535

411-02-02

(b)   have no live part connected to earth or the protective conductor of another system

411-02-05

(c)   are enclosed in an insulating sheath additional to their basic insulation

411-02-06

(d)   have no exposed-conductive-parts connected to earth, to exposed-conductive-parts or protective conductors of other systems or to extraneous-conductive-parts.

411-02-07

PELV
Protective extra-low (PELV) systems must meet all the requirements for SELV, except that the circuits are not electrically separated from earth.

471-14-01
471-14-02

For SELV and PELV systems protection against direct contact need not be provided if voltages do not exceed the following:

411-02-09
471-14-02

Location	SELV	PELV	
Dry areas	25 V a.c. or 60 V d.c.	25 V a.c. or 60 V d.c.	411-02-09
Bathrooms, swimming pools, saunas	Protection required at all voltages	Protection required at all voltages	601-03-02 602-03-01 603-03-01
Other areas	12 V a.c. or 30 V d.c.	6 V a.c. or 15 V d.c.	471-01-01 471-14-02

3.5   Disconnection Times

3.5.1   Conventional Circuits

For the conventional circuits given in Section 7, the correct disconnection time in seconds (0.4 s or 5 s) is obtained by using the protective devices and related maximum circuit lengths in Table 7.1.

3.5.2   Special Circuits

A disconnection time of not more than 0.4 s is required for final circuits supplying:

(i)   portable equipment intended to be moved by hand while in use
413-02-09

(ii)   hand-held metal-cased equipment requiring an earth, and supplied directly or through a socket-outlet
413-02-09

(iii) fixed equipment outside the equipotential zone with accessible exposed-conductive-parts
471-08-03

3.6   Residual Current Devices (RCD)

Note: Residual current device (RCD) is a device type that includes residual current circuit-breakers (RCCBs) and residual current circuit-breakers with integral overcurrent protection (RCBOs).

3.6.1   Protection by an RCD

There are a number of instances where an installation is required to incorporate one or more RCDs (RCCB or RCBO). These instances include:

19

(i)   where the earth fault loop impedance is too high to   413-02-19
       provide the required disconnection time e.g. where the
       supplier does not provide an earth - TT systems

(ii)   on socket-outlet circuits in TT systems   471-08-06

(iii) on all socket-outlets that may reasonably be expected   471-16-01
       to supply portable equipment used outdoors

(iv)   circuits supplying portable equipment for use outdoors   471-16-02
       by means of a flexible cable.

(v)   on socket-outlets in a room, other than a bathroom or   601-08-02
       shower room, containing a shower cubical.

3.6.2 Applications of RCDs

Installations are required to be divided into circuits to avoid   314-01-01
danger and minimise inconvenience in the event of a fault and
to take account of hazards that might arise from the failure of   314-01-02
a single circuit, e.g. a lighting circuit.

30 mA RCDs installed to provide protection to socket-outlets
likely to feed portable equipment outdoors should protect only
those sockets, see Fig 3b.

Where an RCD is fitted only because the earth loop impedance
is too high for shock protection to be provided by an
overcurrent device, for example in a TT system, the rated
residual operating current should not be less than 100 mA.

If two RCDs are installed they should preferably control
separate circuits, see Fig 3a(i), or a time delay 100 mA or
greater RCD (S type) should be installed, see Fig 3a(ii).

The use of RCBOs, see Fig 3a(iii), will minimise inconvenience
in the event of a fault.

The enclosures of RCDs or consumer units incorporating RCDs
in TT installations should be of an all-insulated or Class II
construction. Otherwise, additional precautions recommended
by the manufacturer need to be taken to prevent faults to
earth on the supply side of the RCD.

Fig 3a: Installing RCDs in a TT installation

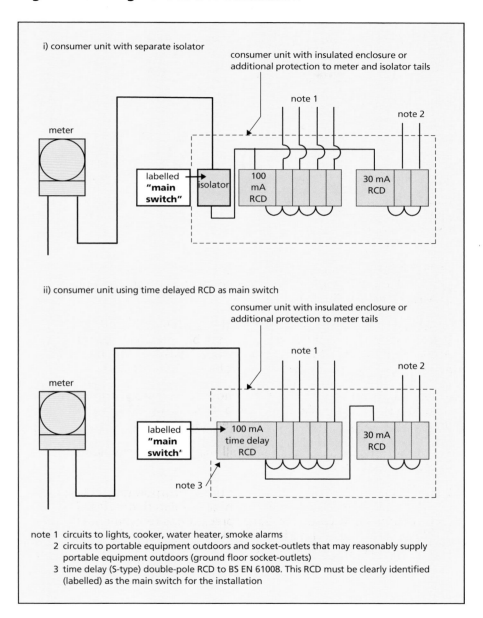

i) consumer unit with separate isolator

consumer unit with insulated enclosure or
additional protection to meter and isolator tails

note 1

note 2

meter

labelled
"main
switch"

isolator

100
mA
RCD

30 mA
RCD

ii) consumer unit using time delayed RCD as main switch

consumer unit with insulated enclosure or
additional protection to meter tails

note 1

note 2

meter

labelled
"main
switch'

100 mA
time delay
RCD

30 mA
RCD

note 3

note 1 circuits to lights, cooker, water heater, smoke alarms
    2 circuits to portable equipment outdoors and socket-outlets that may reasonably supply
       portable equipment outdoors (ground floor socket-outlets)
    3 time delay (S-type) double-pole RCD to BS EN 61008. This RCD must be clearly identified
       (labelled) as the main switch for the installation

21

Fig 3a cont'd: Installing RCDs in a TT installation

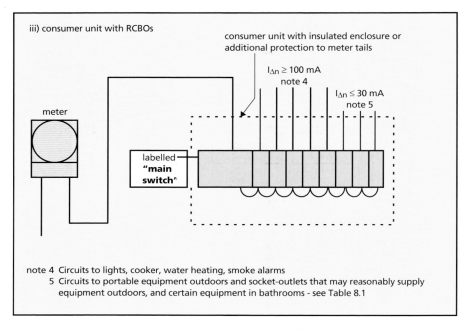

iii) consumer unit with RCBOs

consumer unit with insulated enclosure or additional protection to meter tails

$I_{\Delta n} \geq 100$ mA
note 4

$I_{\Delta n} \leq 30$ mA
note 5

meter

labelled "**main switch**"

note 4 Circuits to lights, cooker, water heating, smoke alarms
5 Circuits to portable equipment outdoors and socket-outlets that may reasonably supply equipment outdoors, and certain equipment in bathrooms - see Table 8.1

3.6.3 Applications of residual current circuit-breakers with overload current protection (RCBOs)

In TN systems it is preferable for reliable operation for indirect shock protection to be provided by overcurrent devices, including RCBOs operating as overcurrent devices; that is, with loop impedances complying with Table 2D of Appendix 2. RCBOs are then providing indirect shock protection as overcurrent devices and supplementary protection against direct contact as residual current circuit-breakers (RCCBs). 413-02-04 471-16

When the designer intends that indirect shock protection is to be provided by a residual current circuit-breaker (RCCB) or the residual current element of an RCBO, loop impedances are as for an RCD, that is appropriate to the rated residual operating current ($Z_s \leq 50$ V / $I_{\Delta n}$), and not more than 200 ohms. 413-02-16

Fig 3b: Installing RCDs in a TN-S or TN-C-S installation

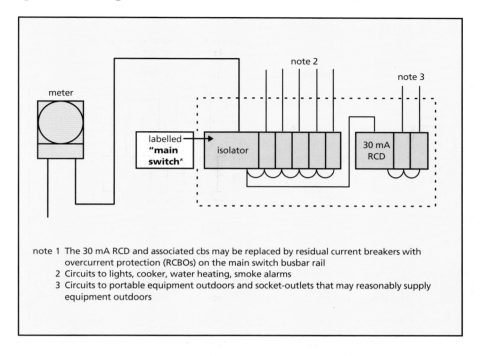

note 1 The 30 mA RCD and associated cbs may be replaced by residual current breakers with
overcurrent protection (RCBOs) on the main switch busbar rail
2 Circuits to lights, cooker, water heating, smoke alarms
3 Circuits to portable equipment outdoors and socket-outlets that may reasonably supply
equipment outdoors

SECTION 4.   BONDING AND EARTHING

*4.1   Main Equipotential Bonding of Metal Services (Figs 4a,
4b, 4c)*

Main equipotential bonding conductors are required to 413-02-02
connect the following metallic parts to the main earthing
terminal, where they are extraneous-conductive-parts:

(i)   metal water service pipes

(ii)   metal gas installation pipes

(iii)   other metal service pipes (including oil and gas supply
pipes) and ducting

(iv)   metal central heating and air conditioning systems

(v)   exposed metallic structural parts of the building

(vi)   lightning protection systems.

*4.2   Main Earthing and Main Equipotential Bonding
        Conductor Cross-Sectional Areas*

The minimum cross-sectional area (csa) of the main equipotential 547-02-01
bonding conductor is half that of the main earthing
conductor. For 100 A TN installations, the main earthing
conductor csa needs to be 16 mm^2 and that of the main
bonding conductors 10 mm^2 where the size of the supply Table 54H
neutral conductor is not more than 35 mm^2. However, local
public electricity distribution network conditions may require
larger conductors. For other conditions see Table 10A of
Appendix 10. For TT installations see Fig 4c.

Note that:

(i)   only copper conductors should be used; copper covered 542-03-03
        aluminium conductors or aluminium conductors or
        structural steel can only be used if special precautions
        outside the scope of this Guide are taken

(ii)   bonding connections to incoming metal services should 547-02-02
        be as near as possible to the point of entry of the
        services to the premises, but on the consumer's side of
        any insulating section

(iii)   the connection to the gas, water, oil, etc service should 547-02-02
        be within 600 mm of the service meter, or at the point
        of entry to the building if the service meter is external,
        and must be on the consumer's side before any branch
        pipework and after any insulating section in the service.
        The connection must be made to hard pipe, not to soft
        or flexible meter connections

(iv)   the connection must be made using clamps (to BS 951) 542-03-03
        which will not be subject to corrosion at the point of
        contact

(v)   if incoming gas and water services are of plastic, main
        bonding connections are to be made to metal
        installation pipes only.

4.3   Main Equipotential Bonding - Plastic Services

There is no requirement to main bond an incoming service where both the incoming service pipe and the pipework within the installation are both of plastic. Where there is a plastic incoming service and a metal installation within the premises, main bonding must be carried out, the bonding being applied on the customer's side of any meter, main stop cock or insulating insert.

4.4   Earthing

Every exposed-conductive-part (a part which may become live 413-02-06 under earth fault conditions) shall be connected by a protective 413-02-18 conductor to the main earthing terminal.

4.5   Supplementary Equipotential Bonding in Locations of Increased Shock Risk - Metal Pipework

Supplementary equipotential bonding is required only in locations of increased shock risk such as some of those in Part 6 of BS 7671 (471-08-01). In domestic premises, the locations 471-08-01 identified as having increased shock risks are rooms containing a bath or shower (bathrooms) and around swimming pools.

In a bathroom or shower room, local supplementary equipotential 601-04-01 bonding is required to be provided connecting together the terminal of protective conductors of each circuit supplying Class I and Class II equipment in zones 1, 2 or 3, and extraneous-conductive-parts in these zones including the following:

(i)   metal pipes supplying services and metallic waste pipes (e.g. water, gas)

(ii)   metal central heating pipes and air conditioning systems

(iii)   accessible metal structural parts of the building (metal door architraves, metal handrails, window frames and similar parts are not considered to be extraneous-conductive-parts unless they are connected to metallic structural parts of the building)

(iv)   metal baths and metal shower basins.

Circuit protective conductors may be used as supplementary bonding conductors.

25

The supplementary equipotential bonding may be provided in close proximity to the location.

See Section 8.1 for locations containing a bath or shower.

A typical installation is shown in Figure 4d.

601-04-01

4.6 Supplementary Bonding in other Locations - Metal Pipework

471-08-01

There is no specific requirement in BS 7671 to supplementary bond the following

kitchen pipes, sinks or draining boards

metal furniture in kitchens

metal pipes and wash hand basins in domestic locations other than bathrooms.

Note: Metal waste pipes in contact with earth should be main bonded back to the main earthing terminal.

4.7 Supplementary Bonding of Plastic Pipe Installations

Supplementary bonding is not required to metal parts supplied by plastic pipes, such as metal hot and cold water taps supplied from plastic pipes. A metal bath not connected to extraneous-conductive-parts (such as structural steelwork) with plastic hot and cold water pipes and plastic waste pipes does not require supplementary bonding. Supplementary bonding in a bathroom or shower room will still be required between the protective conductors of circuits supplying Class I and Class II equipment in the zones e.g. heaters, showers and accessible luminaires, see Figure 4e.

Fig 4a: Typical earthing arrangements and protective conductor csa - TN-S

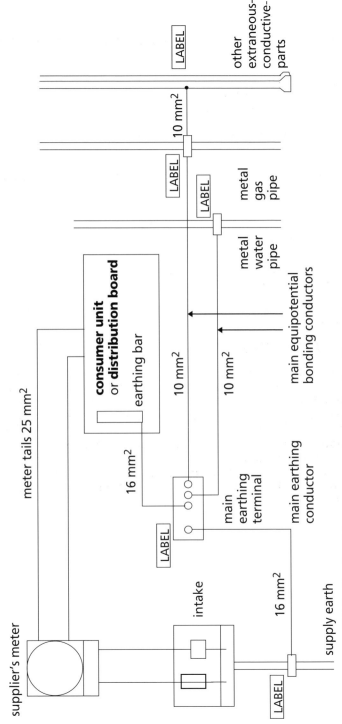

supplier's meter

meter tails 25 mm^2

consumer unit
or **distribution board**

earthing bar

16 mm^2

10 mm^2

10 mm^2

LABEL

LABEL

LABEL

10 mm^2

LABEL

other extraneous-conductive-parts

metal gas pipe

metal water pipe

main equipotential bonding conductors

main earthing terminal

main earthing conductor

intake

16 mm^2

supply earth

TN-S: Earthed to armour or metallic sheath

Note: i) Main equipotential bonding conductors may be separate (as shown) or looped with unbroken conductors.

ii) LABEL - Safety Electrical Connection - Do Not Remove.

27

Fig 4b: Typical earthing arrangements and protective conductor csa - TN-C-S

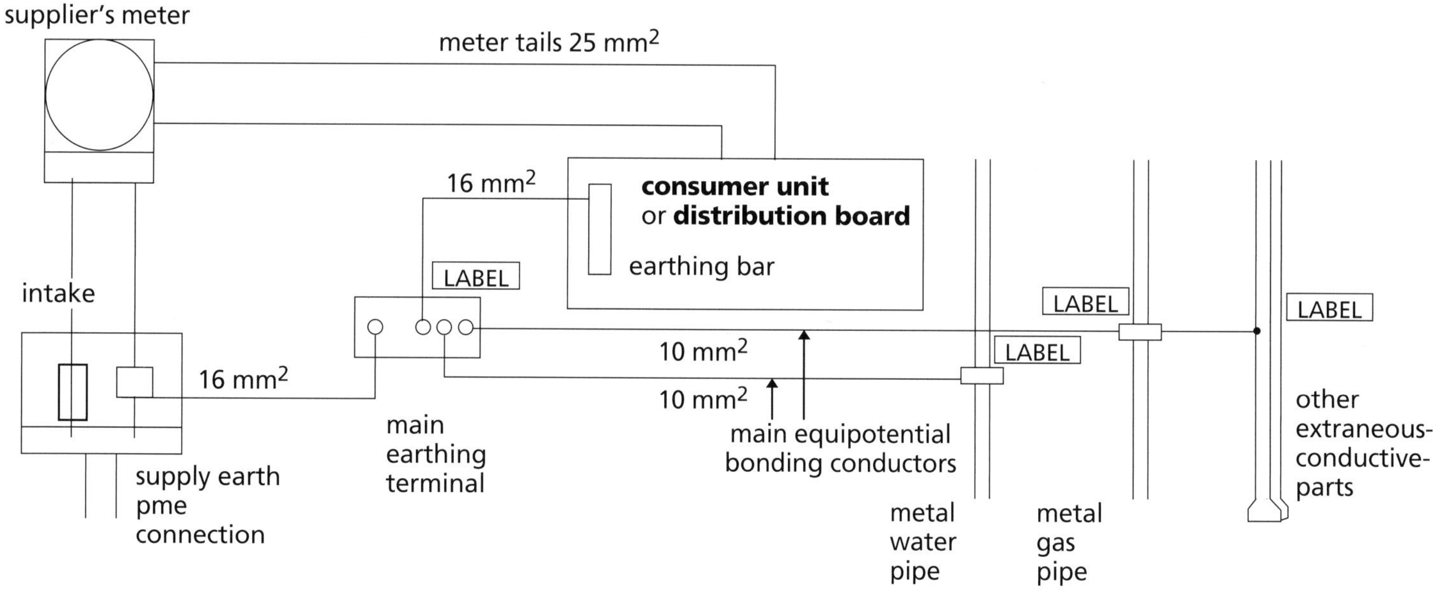

supplier's meter

meter tails 25 mm²

consumer unit or **distribution board**

earthing bar

16 mm²

LABEL

main earthing terminal

16 mm²

supply earth pme connection

intake

10 mm²

10 mm²

main equipotential bonding conductors

metal water pipe

metal gas pipe

LABEL

LABEL

LABEL

other extraneous-conductive-parts

TN-C-S: PME earth

Note: i) Main equipotential bonding conductors may be separate (as shown) or looped with unbroken conductors.

ii) Local electricity supply network conditions may require larger conductors.

iii) LABEL - Safety Electrical Connection - Do Not Remove.

28

Fig 4c: Typical earthing arrangements and protective conductor csa - TT

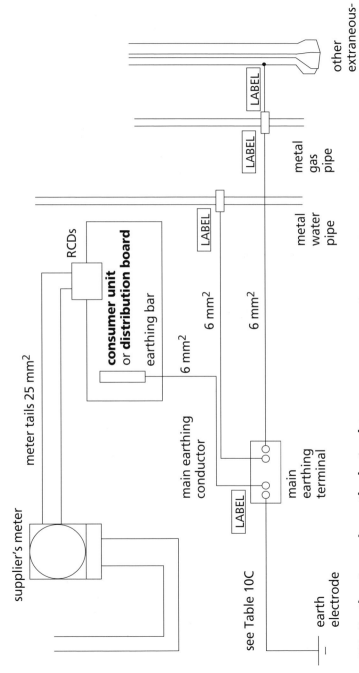

TT: Earth return via earth electrode

Note: i) Main equipotential bonding conductors may be separate (as shown) or looped with unbroken conductors.

ii) LABEL - Safety Electrical Connection - Do Not Remove.

iii) An earth electrode resistance exceeding 1Ω is presumed, see paragraph 4.10.

Fig 4d: Supplementary bonding in a bathroom - metal pipe installation

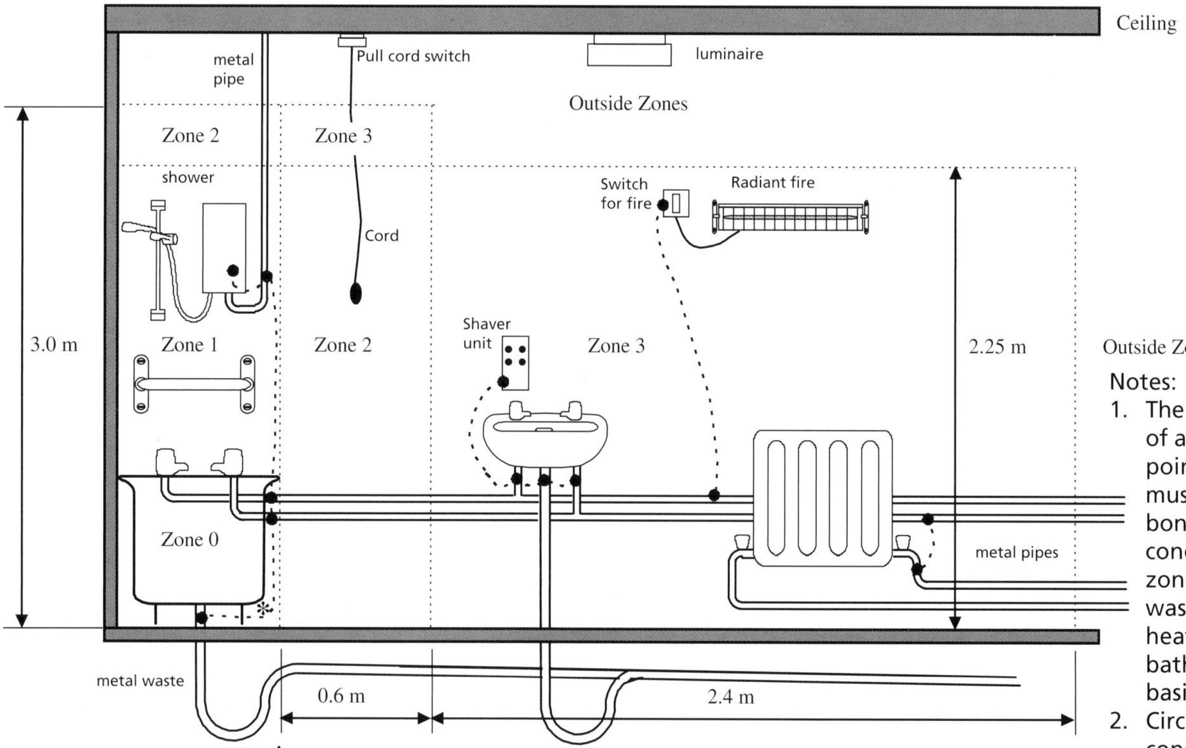

Ceiling

luminaire

Outside Zones

Radiant fire

Switch for fire

Zone 3

Shaver unit

Pull cord switch

Zone 3

Cord

Zone 2

metal pipe

Zone 2

shower

Zone 1

Zone 0

metal waste

3.0 m

Outside Zones

2.25 m

metal pipes

2.4 m

0.6 m

* Zone 1 if the space is accessible without the use of a tool.
Spaces under the bath, accessible only with the use of a tool, are outside the zones.

Notes:

1. The protective conductors of all power and lighting points within the zones must be supplementary bonded to all extraneous-conductive-parts in the zones, including metal waste, water and central heating pipes, and metal baths and metal shower basins.

2. Circuit protective conductors may be used as supplementary bonding conductors.

Fig 4e: Supplementary bonding in a bathroom - plastic pipe installation

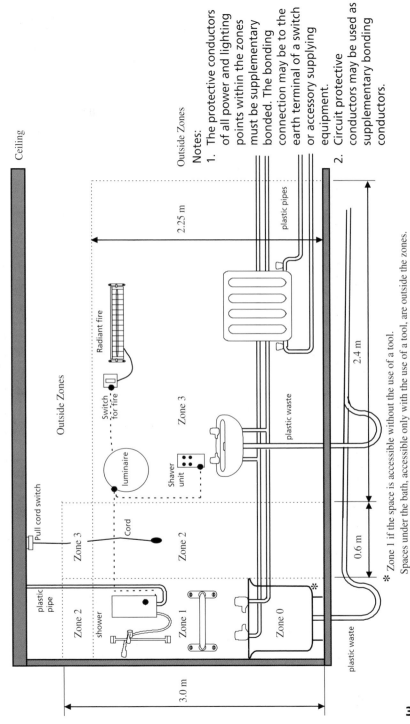

Notes:

1. The protective conductors of all power and lighting points within the zones must be supplementary bonded. The bonding connection may be to the earth terminal of a switch or accessory supplying equipment.

2. Circuit protective conductors may be used as supplementary bonding conductors.

* Zone 1 if the space is accessible without the use of a tool.
Spaces under the bath, accessible only with the use of a tool, are outside the zones.

4.8 Earth Electrode (Fig 4c)

This is connected to the main earthing terminal by the earthing 542-01-04
conductor and provides part of the earth fault loop for a TT
installation.

It is recommended that the earth fault loop impedance for TT 542-02-02
installations does not exceed 200 ohms.

Metal gas or water or other metal service pipes are not to be 542-02-04
used as the earth electrode, although they must be bonded as
paragraph 4.1.

4.9 Types of Earth Electrode

The following types of earth electrode are recognised:

(i)   earth rods or pipes   542-02-01

(ii)   earth tapes or wires   542-02-01

(iii)   earth plates   542-02-01

(iv)   underground structural metalwork embedded in   542-02-01
        foundations

(v)   welded metal reinforcement of concrete embedded in   542-02-01
        the earth (excluding pre-stressed concrete)

(vi)   lead sheaths and metal coverings of cables, which must   542-02-05
        meet the following conditions:

        (a)   the sheath or covering shall be in effective contact   542-02-05(ii)
              with earth,

        (b)   the consent of the owner of the cable shall be   542-02-05(iii)
              obtained, and

        (c)   arrangements shall be made for the owner of   542-02-05(iv)
              the cable to warn the owner of the electrical
              installation of any proposed change to the cable
              or its method of installation which might affect
              its suitability as an earth electrode.

4.10 Typical Earthing Arrangements for Various Types of Earthing System (Figs 4a, 4b, 4c)

Figs 4a, 4b and 4c show the single-phase arrangements, but three-phase arrangements are similar.

The protective conductor sizes shown in Figures refer to copper conductors and are related to 25 mm² supply tails from the meter.

Table 54G
Table 54H
547-02-01

For TT systems protected by an RCD with an earth electrode resistance 1 ohm or greater, the earthing conductor size need not exceed 2.5 mm² if protected against corrosion by a sheath and if also protected against mechanical damage; otherwise, see Table 10C of Appendix 10.

542-03-01
543-01-03

The earthing bar is sometimes used as the main earthing terminal; however, means must be provided in an accessible position for disconnecting the earthing conductor to facilitate testing of the earthing.

542-04-02

Note: For TN-S and TN-C-S installations, advice about the availability of an earthing facility and the precise arrangements for connection should be obtained from the electricity supplier.

SECTION 5. ISOLATION AND SWITCHING

5.1   Isolation

A means of isolation must be provided to enable electrically 460-01-01
skilled persons to carry out work on, or near, parts which
would otherwise normally be energised. Isolating devices must
comply with the isolation requirements of BS 1363-4, BS 3676, 511
BS EN 60669-2-4, BS EN 60898, BS EN 60947-2, BS EN 60947-3, 537-02-02
BS EN 61008-1 or BS EN 61009-1. The position of the contacts 537-02-04
must either be externally visible or be clearly, positively and 476-02-02
reliably indicated. If it is installed remotely from the
equipment to be isolated, the device must be capable of being
secured in the OPEN position.

Means of isolation should be provided as follows:

(i)   at the origin of the installation, a main linked switch or 460-01-02
         circuit-breaker should be provided as a means of
         isolation and of interrupting the supply on load. For
         single-phase household and similar supplies that may 476-01-03
         be operated by unskilled persons, a double-pole device
         must be used for both TT and TN systems. For
         three-phase supplies an isolator must interrupt the phase 460-01-04
         and neutral conductors in a TT system; in a TN-S or TN-C-S
         system only phase conductors need be interrupted.
         Provision shall be made for disconnecting the neutral 460-01-06
         conductor. Where this is a joint it shall be such that it is
         accessible, can only be disconnected by means of a tool,
         is mechanically strong and will reliably maintain
         electrical continuity

(ii)   other than at the origin of the installation every circuit, 476-01-02
         or group of circuits, which may have to be isolated
         without interrupting the supply to other circuits should
         be provided with its own isolating device, which must 537-02-01
         switch all live conductors in a TT system and all phase
         conductors in a TN system

Every motor circuit should be provided with a readily accessible 131-14-02
device to switch off the motor and all associated equipment
including any automatic circuit-breaker.

5.2　Functional Switching

537-05

A means of switching for interrupting the supply on load is 476-01-02 required for every circuit and final circuit.

One common switch may be used to interrupt the supply to a 476-01-01 group of circuits. Additionally, a separate switch must be provided for every circuit which, for safety reasons, has to be 476-01-02 switched independently.

5.3　Switching Off for Mechanical Maintenance

462

A means of switching off for mechanical maintenance is 462-01-01 required where mechanical maintenance may involve a risk of burns or of injury from mechanical movement, and may be required for lamp replacement.

Each device for switching off for mechanical maintenance must:

(i)　be capable of switching full load current 537-03-04

(ii)　be suitably located in a readily accessible position 462-01-02

(iii)　be identified with a permanent label unless its purpose 462-01-02 is obvious

(iv)　have either an externally visible contact gap or a 537-03-02 clearly and reliably indicated OFF position. An indicating light should not be relied upon

(v)　be selected and installed to prevent unintentional 537-03-03 reclosure, such as might be caused by mechanical shock or vibration.

5.4　Emergency Switching

463

An emergency switch is to be provided for every part of an 476-01-01 installation which may have to be disconnected rapidly from 476-03-04 the supply to prevent or remove danger. Where there is a risk of electric shock the emergency switch is to disconnect all live 463-01-01 conductors, except in three-phase TN-S and TN-C-S systems 460-01-04 where the neutral need not be switched.

A means of emergency stopping is also to be provided where 463-01-05 mechanical movement of electrically actuated equipment may 476-03-02 give rise to danger.

A plug and socket-outlet or similar device shall not be selected 537-04-02 as a device for emergency switching.

An emergency switch must be:

(i)   readily accessible from the place where the danger   537-04-04
        may occur

(ii)   marked, preferably with a red handle or pushbutton   537-04-04

(iii)   capable of cutting off the full load current   537-04-01

(iv)   of the latching type or capable of being restrained in 537-04-05
        the 'OFF' or 'STOP' position

(v)   double-pole for single-phase systems   463-01-01

(vi)   if operated via a relay or contactor, of a design which 537-04-03
        has fail-safe characteristics.

A fireman's switch must be provided to disconnect the supply to 476-03-05 a high voltage installation, e.g. a neon sign, but such installations are outside the scope of this Guide (see Regulations 476-03-05 to 07 and 537-04-06 of BS 7671).   537-04-06

SECTION 6. LABELLING

6.1   Labels to be Provided

The following durable labels are to be securely fixed on or adjacent to equipment installed in final circuits.

(i)   *Unexpected presence of nominal voltage (U or U_o)*   514-10-01
        exceeding 230 V

        Where the nominal voltage (U or U_o) exceeds 230 V,
        e.g. 400 V phase-to-phase, and it would not normally be
        expected to be so high, a warning label stating the
        maximum voltage present shall be provided where it
        can be seen before gaining access to live parts.

(ii) *Nominal voltage exceeding 230 volts (U or U_o)* 514-10-01
between simultaneously accessible equipment

For simultaneously accessible equipment with terminals or other fixed live parts having a nominal voltage (U or U_o) exceeding 230 volts between them, e.g. 400 V phase-to-phase, a warning label shall be provided where it can be seen before gaining access to live parts.

(iii) *Presence of different nominal voltages in the same* 514-10-01
equipment

Where equipment contains different nominal voltages, e.g. both low and extra-low, a warning label stating the voltages present shall be provided so that it can be seen before gaining access to simultaneously accessible live parts.

(iv) *Connection of earthing and bonding conductors* 514-13

A label to BS 951 durably marked with the words as 514-13-01
follows

SAFETY ELECTRICAL CONNECTION - DO NOT REMOVE

shall be permanently fixed in a visible position at or near the point of connection of

(1) every earthing conductor to an earth electrode or other means of earthing, and

(2) every bonding conductor to extraneous-conductive-parts, and

(3) at the main earth terminal, where it is not part of the main switchgear.

(v) *Purpose of switchgear and controlgear* 514-01-01

Unless there is no possibility of confusion, a label indicating the purpose of each item of switchgear and controlgear shall be fixed on or adjacent to the gear. It may be necessary to label the item controlled, as well as its controlgear.

Identification of protective devices   514-08-01

A protective device, e.g. fuse or circuit-breaker, shall be arranged and labelled so that the circuit protected may be easily recognised.

(vii)   *Identification of isolators*   461-01-05
                                                537-02-09

All isolating devices shall be clearly and durably marked to indicate the circuit or circuits which they isolate.

(viii) *Isolation requiring more than one device*   514-11-01

A durable warning notice must be permanently fixed in a clearly visible position to identify the appropriate isolating devices, where equipment or an enclosure contains live parts which cannot be isolated by a single device.

(ix)   *Periodic inspection and testing*   514-12-01

A notice of durable material indelibly marked with the words as follows, and no smaller than the example shown in BS 7671,

---

IMPORTANT

This installation should be periodically inspected and tested and a report on its condition obtained, as prescribed in BS 7671 Requirements for Electrical Installations published by the Institution of Electrical Engineers.

Date of last inspection ....

Recommended date of next inspection ....

---

shall be fixed in a prominent position at the origin of every installation. The electrician carrying out the initial verification must complete the notice, and it must be updated after each periodic inspection.

(x)   *Diagrams*

A diagram, chart or schedule shall be provided showing:

(a)   the number of points, size and type of cables for each circuit

(b)   the method of providing protection against indirect contact

(c)   any circuit vulnerable to an insulation test.

The schedules of test results (Form F4) of Appendix 7 meets the above requirement for a schedule.

(xi)   *Residual current devices*

Where an installation incorporates an RCD a notice with the words as follows, and no smaller than the example shown in BS 7671,

---

This installation, or part of it, is protected by a device which automatically switches off the supply if an earth fault develops. Test quarterly by pressing the button marked 'T' or 'Test'. The device should switch off the supply and should then be switched on to restore the supply. If the device does not switch off the supply when the button is pressed, seek expert advice.

---

shall be fixed in a prominent position at or near the origin of the installation.

SECTION 7. FINAL CIRCUITS

7.1   Final Circuits

413-02-08
413-02-12
525-01-02
543-01-03
App 4
Table 4D2A
Table 4D2B

Table 7.1 has been designed to enable a radial or ring final circuit to be installed without calculation where the supply is at 230 V single-phase or 400 V three-phase. For other voltages, the maximum circuit length given in the table must be corrected by the application of the formula

$$L_p = \frac{L_t \times U_o}{230}$$

where:

L_p is the permitted length for voltage U_o,
L_t is the tabulated length for 230 V
U_o is the supply voltage.

The conditions assumed are that:

(i)   the installation is supplied

    (a)   by a TN-C-S system with a maximum external earth fault loop impedance, Z_e, of 0.35 ohm, or

    (b)   by a TN-S system with a maximum Z_e of 0.8 ohm, or

    (c)   a TT system with RCDs installed as described in Section 3.6.

(ii)   the final circuit is connected to a distribution board or consumer unit at the origin of the installation

(iii)   the method of installation complies with Reference Methods 1, 3 or 6 of Appendix 4 of BS 7671: App 4

    (a)   Reference Method 1 (M1)
        Sheathed cables, armoured or unarmoured clipped direct or embedded in plaster

    (b)   Reference Method 3 (M3)
        Cables run in conduit or trunking
        Single-core or insulated and sheathed

(c)   Reference Method 6 (M6)

Multicore thermoplastic (pvc) insulated and sheathed   Table 4D5A
flat twin cable with protective conductor in conduit in
an insulating wall and also direct in a thermally
insulating wall, or single-core thermoplastic (pvc)
insulated cables enclosed in conduit in a thermally
insulating wall (Method 15).

(iv)   the ambient temperature throughout the length of the   Table 4C1
circuit does not exceed 30 °C

(v)   the characteristics of protective devices are in   App 3
accordance with Appendix 3 of BS 7671, with a fault
current tripping time for circuit-breakers of 0.1 s or less

(vi)   the cable conductors are of copper.

7.2   *Using the Tabulated Final Circuits*

7.2.1   *Grouping of Cables*

(i)   In domestic premises, except for heating cables, the   Table 4B1
conventional circuit design permits any number of
single-layer circuits when the spacing between
adjacent surfaces of the cables exceeds one cable
diameter, and, for other than semi-enclosed fuses,
(BS 3036) up to 5 touching, single-layer, circuits, when
clipped to a non-metallic surface (Installation Method 1)

(ii)   up to four, 5 A or 6 A circuits of enclosed or bunched   Table 4B1
cables (Methods 3 and 6) are allowed for circuits
protected by semi-enclosed fuses to BS 3036 and up to
6 circuits when protected by BS 88 or BS 1361 fuses or
by circuit-breakers to BS 3871-1 or BS EN 60898 or
RCBOs to BS EN 61009

(iii)   for other groupings and/or high ambient temperatures
and/or enclosure in thermal insulation cable sizes will
need to be increased per Appendix 6 of this Guide.

41

TABLE 7.1 Conventional Circuits

Device rating A	Cable size mm² phase/cpc	Protective device type (note v)	Cable Installation method (note i) thermo-plastic (pvc) cable	thermo-setting cable	Maximum length in metres $Z_e \leq 0.8$ ohm TN-S 0.4 s disconnection	5 s disconnection	$Z_e \leq 0.35$ ohm TN-C-S 0.4 s disconnection	5 s disconnection
Ring Circuits								
30	2.5/1.5	BS 1361	M6	M6	90(iii)	N1	90	N1
		BS 3036	M1	M3	91(iv)	N1	91	N1
		cb Type 2	M6	M6	58(ii)	N1	88	N1
		cb/RCBO Type 1 & B	M6	M6	88	N1	88	N1
		cb/RCBO Type 3 & C	M6	M6	N2	N1	76(ii)	N1
32	2.5/1.5	BS 88	M6	M6	66(iii)	N1	66	N1
		cb/RCBO Type 1 & B	M6	M6	84	N1	84	N1
		cb Type 2	M6	M6	46(ii)	N1	84	N1
		cb/RCBO Type 3 & C	M6	M6	N2	N1	68(ii)	N1
Radial Circuits								
5	1.0/1.0	BS 1361, BS 3036	M6	M6	46	46	46	46
		cb/RCBO Type 1, 2, 3, B & C	M6	M6	46	46	46	46
5	1.5/1.0	BS 1361, BS 3036	M6	M6	71	71	71	71
		cb/RCBO Type 1, 2, 3, B & C	M6	M6	71	71	71	71
6	1.0/1.0	BS 88	M6	M6	38	38	38	38
		cb/RCBO Type 1, 2, 3, B & C	M6	M6	38	38	38	38
		cb/RCBO Type D	M6	M6	27(ii)	27(ii)	38(ii)	38(ii)

TABLE 7.1 continued **Conventional Circuits**

Device rating A	Cable size mm² phase/cpc	Protective device type (note v)	Cable Installation method (note i)		Maximum length in metres			
					$Z_e \leq 0.8$ ohm TN-S		$Z_e \leq 0.35$ ohm TN-C-S	
			thermo-plastic (pvc) cable	thermo-setting cable	0.4 s disconnection	5 s disconnection	0.4 s disconnection	5 s disconnection
Radial Circuits								
6	1.5/1.0	BS 88	M6	M6	59	59	59	59
		cb/RCBO Type 1, 2, 3, B & C	M6	M6	59	59	59	59
		cb/RCBO Type D	M6	M6	33(ii)	33(ii)	45(ii)	45(ii)
10	1.0/1.0	BS 88	M6	M6	21	21	21	21
		cb/RCBO Type 1, 2, 3, B & C	M6	M6	21	21	21	21
		cb/RCBO Type D	M6	M6	9(ii)	9(ii)	20(ii)	20(ii)
10	1.5/1.0	BS 88	M6	M6	33	33	33	33
		cb/RCBO Type 1, 2, 3, B & C	M6	M6	33	33	33	33
		cb/RCBO Type D	M6	M6	11(ii)	11(ii)	23(ii)	23(ii)
15	2.5/1.5	BS 3036	M3	M6	35	35	35	35
		cb/RCBO Type 1, 2 & B	M6	M6	35	35	35	35
		cb/RCBO Type 3 & C	M6	M6	34(ii)	34(ii)	35	35
15	4.0/1.5	BS 3036	M6	M6	61	61	61	61
		BS 1361 cb/RCBO Type 1, 2 & B	M6	M6	61	61	61	61
		cb/RCBO Type 3 & C	M6	M6	39(ii)	39(ii)	61(ii)	61(ii)
16	2.5/1.5	BS 88	M6	M6	33	33	33	33
		BS 1361 cb/RCBO Type 1, 2 & B	M6	M6	33	33	33	33
		cb/RCBO Type 3 & C	M6	M6	29(ii)	29(ii)	33	33
		cb/RCBO Type D	M6	M6	N2	N2	17(ii)	17(ii)

TABLE 7.1 continued **Conventional Circuits**

Device rating A	Cable size mm² phase/cpc	Protective device type (note v)	Cable Installation method (note i) thermoplastic (pvc) cable	thermosetting cable	Maximum length in metres $Z_e \leq 0.8$ ohm TN-S 0.4 s disconnection	5 s disconnection	$Z_e \leq 0.35$ ohm TN-C-S 0.4 s disconnection	5 s disconnection
Radial Circuits								
16	4.0/1.5	BS 88	M6	M6	56	56	56	56
		cb/RCBO Type 1, 2 & B	M6	M6	56	56	56	56
		cb/RCBO Type 3 & C	M6	M6	34(ii)	34(ii)	56(ii)	46(ii)
		cb/RCBO Type D	M6	M6	N2	N2	20(ii)	20(ii)
20	2.5/1.5	BS 88, BS 1361	M3	M6	27	27	27	27
		BS 3036	M3	M3	N3	N3	N3	N3
		cb/RCBO Type 1, 2 & B	M3	M6	27	27	27	27
		cb/RCBO Type 3 & C	M3	M6	17(ii)	17(ii)	27	27
		cb/RCBO Type D	M6	M6	N2	N2	10(ii)	10(ii)
20	4.0/1.5	BS 3036	M3	M6	43	43	43	43
		BS 88, BS 1361	M6	M6	43	43	43	43
		cb/RCBO Type 1 & B	M6	M6	43	43	43	43
		cb Type 2	M6	M6	43	43	43	43
		cb/RCBO Type 3 & C	M6	M6	19	19	43(ii)	42(ii)
		cb/RCBO Type D	M6	M6	N2	N2	12	12
25	4.0/2.5	BS 88	M6	M6	33	33	33	33
		cb/RCBO Type 1, 2 & B	M6	M6	33	33	33	33
		cb/RCBO Type 3 & C	M6	M6	11(ii)	11(ii)	33	33
		cb/RCBO Type D	M6	M6	N2	N2	9(ii)	9(ii)

TABLE 7.1 continued **Conventional Circuits**

Device rating A	Cable size mm² phase/cpc	Protective device type (note v)	Cable Installation method (note i)		Maximum length in metres			
					Ze ≤ 0.8 ohm TN-S		Ze ≤ 0.35 ohm TN-C-S	
			thermo-plastic (pvc) cable	thermo-setting cable	0.4 s disconnection	5 s disconnection	0.4 s disconnection	5 s disconnection
Radial Circuits								
25	4.0/1.5	BS 88	M6	M6	33	33	33	33
		cb/RCBO Type 1 & B	M6	M6	33	33	33	33
		cb Type 2	M6	M6	28(ii)	28(ii)	33	33
		cb/RCBO Type 3 & C	M6	M6	8(ii)	8(ii)	30(ii)	30(ii)
		cb/RCBO Type D	M6	M6	N2	N2	N2	N2
30	6.0/2.5	BS 3036	M1	M6	27(ii)	45	45	45
		BS 1361	M6	M6	31(ii)	42	42	42
		cb/RCBO Type 1 & B	M6	M6	42	42	42	42
		cb Type 2	M6	M6	27(ii)	27(ii)	42	42
		cb/RCBO Type 3 & C	M6	M6	N2	N2	35(ii)	35(ii)
30	10.0/4.0	BS 3036	M6	M6	44(ii)	74	74	74
		BS 1361	M6	M6	51(ii)	74	74	74
		cb/RCBO Type 1 & B	M6	M6	74	74	74	74
		cb Type 2	M6	M6	44(ii)	44(ii)	74	74
		cb/RCBO Type 3 & C	M6	M6	N2	N2	58(ii)	58(ii)
32	6.0/2.5	BS 88	M6	M6	23(ii)	39	39	39
		cb/RCBO Type 1 & B	M6	M6	39	39	39	39
		cb Type 2	M6	M6	21(i)	21(i)	39	39
		cb/RCBO Type 3 & C	M6	M6	N2	N2	31(ii)	31(ii)
		cb/RCBO Type D	M6	M6	N2	N2	2(ii)	2(ii)

TABLE 7.1 continued **Conventional Circuits**

Device rating A	Cable size mm² phase/cpc	Protective device type (note v)	Cable Installation method (note i)		Maximum length in metres			
					$Z_e \leq 0.8$ ohm TN-S		$Z_e \leq 0.35$ ohm TN-C-S	
			thermo-plastic (pvc) cable	thermo-setting cable	0.4 s disconnection	5 s disconnection	0.4 s disconnection	5 s disconnection
Radial Circuits 32	10.0/4.0	BS 88	M6	M6	37(ii)	69	69	69
		cb/RCBO Type 1 & B	M6	M6	69	69	69	69
		cb Type 2	M6	M6	35(ii)	35(ii)	69	69
		cb/RCBO Type 3 & C	M6	M6	N2	N2	51(ii)	51(ii)
40	10.0/4.0	BS 88	M6	M6	7(ii)	53	53	53
		cb Type 1	M6	M6	53	53	53	53
		cb/RCBO Type B	M6	M6	51(ii)	51(ii)	53	53
		cb Type 2	M6	M6	7(ii)	7(ii)	53	53
		cb/RCBO Type 3 & C	M6	M6	N2	N2	32(ii)	32(ii)
40	16.0/6.0	BS 88	M6	M6	11(ii)	88	88	88
		cb Type 1	M6	M6	88	88	88	88
		cb/RCBO Type B	M6	M6	78(ii)	78(ii)	88	88
		cb Type 2	M6	M6	11(ii)	11(ii)	88	88
		cb/RCBO Type 3 & C	M6	M6	N2	N2	49(ii)	49(ii)
45	10/4.0	BS 1361	M3	M6	N2	10(ii)	32(ii)	49
		BS 3036	M1	M3	N2	49	34(ii)	49
		cb/RCBO Type 1 & B	M3	M6	49	49	49	49
		cb/RCBO Type B	M3	M6	35(i)	35(i)	49	49
		cb Type 2	M3	M6	N2	N2	49	49
		cb/RCBO Type 3 & C	M3	M6	N2	N2	23(i)	23(i)

Notes to Table 7.1

(i) Installation reference method
M6 indicates methods of cable installation M1, M3 and M6 may be used
M3 indicates methods of cable installation M1 and M3 may be used
M1 indicates method of cable installation M1 only may be used

All the circuits are limited by voltage drop other than those marked as below:

(ii) Length is limited by earth fault loop impedance

(iii) Alternative method of Regulation 413-02-12 applied,   413-02-12
$R_2 \div 4$ to be less than 0.29 ohm

(iv) Alternative method of Regulation 413-02-12 applied,   413-02-12
$R_2 \div 4$ to be less than 0.34 ohm

(v) Application of RCBOs, see 3.6.3

N1 NOT PERMISSIBLE as 0.4 s disconnection required for socket-outlet circuits

N2 NOT PERMISSIBLE as earth fault loop impedance too high

N3 NOT PERMISSIBLE cable overloaded

If the alternative method ((iii) or (iv)) is used, R_2 must be recorded on the installation schedule.

Reference to BS 88 fuses is to BS 88-2.1 or BS 88-6
BS 88 fuses are not available in Consumer Units.

7.2.2 Thermosetting cables (e.g. to BS 5467 or BS 7211) Table 4A2

Cable sizes must not be reduced when cables with thermosetting insulation are used, as the cable operating temperature may exceed the maximum tolerated by the accessory to which it is connected.

7.2.3 Checklist

Before installing a conventional final circuit the following questions must be answered:

(i)   what is the load current and can the distribution board and supply arrangements accommodate it? (See Appendix 1 for guidance on assessing load currents)

(ii)   which kind of protective device is to be used?

(iii)   what cable type and installation method are to be used?

(iv)   what rating of the protective device is equal to or next higher than the load current of the circuit?

(v)   which type of earthing arrangement is employed?

(vi)   is the maximum required disconnection time 0.4 s or 5 s? Maximum 0.4 s disconnection time is required for circuits feeding socket-outlets and circuits feeding fixed equipment outside the equipotential zone. 413-02-09
471-08-03

(vii)   what are the isolation and switching requirements? (See Section 5)

(viii) what labels are required? (See Section 6)

(ix)   is the earth loop impedance value below the values given in 7.1(i) or 7.2.4(ii)?

(x)   is an RCD or RCBO required? All socket-outlets on a TT system must be protected by an RCD or RCBO. Socket-outlets in all systems, that may reasonably be expected to supply equipment outdoors need to be protected by an RCD or RCBO with a rated residual operating current of 30 mA (all RCDs or RCBOs to comply with BS 4293, BS 7288, BS EN 61008 or BS EN 61009). 471-08-06

471-16-01

Certain equipment in bathrooms requires RCD protection. 601-09-02
601-09-03

7.2.4   TT Systems

For TT systems the figures for TN-C-S systems, with 5 s disconnection time, may be used provided that:

(i)   the circuit is controlled by an RCD to BS 4293, BS EN 61008 or BS EN 61009 with a rated residual operating current not exceeding 200 mA, and

(ii)   the total earth fault loop impedance is verified as being less than 200 ohms, and

(iii)   a device giving both overload and short-circuit protection is installed in the circuit. This may be an RCBO.

7.2.5   Choice of Protective Device

The selection of protective device depends upon:

(i)   prospective fault current
(ii)   circuit load characteristics
(iii)   cable current-carrying capacity
(iv)   disconnection time limit.

Whilst these factors have generally been allowed for in the conventional final circuits in Table 7.1, the following additional guidance is given:

(i)   *prospective fault current*

If a protective device is to operate safely its rated short-circuit capacity must exceed the prospective fault current at the point it is installed.   434-03-01

At the origin of the installation the supplier needs to be consulted as to the prospective fault current. Except for London and some other major city centres, the maximum fault current for 230 V single-phase supplies up to 100 A will not exceed 16 kA.   313-01-01

Consumer units including protective devices complying as a whole assembly with BS 5486-13 or BS EN 60439-3 are suitable for locations with fault currents up to 16 kA when supplied through a type II fuse to BS 1361 : 1971 (1992) rated at no more than 100 A.

TABLE 7.2A

Rated Short-Circuit Capacities

Device type	Device designation	Rated short-circuit capacity kA	
Semi-enclosed fuse to BS 3036 with category of duty	S1A S2A S4A	1 2 4	
Cartridge fuse to BS 1361 type I type II		16.5 33.0	
General purpose fuse to BS 88-2.1		50 at 415 V	
General purpose fuse to BS 88-6		16.5 at 240 V 80 at 415 V	
Circuit-breakers to BS 3871 (replaced by BS EN 60898)	M1 M1.5 M3 M4.5 M6 M9	1 1.5 3 4.5 6 9	
Circuit-breakers to BS EN 60898* and RCBOs to BS EN 61009		I$_{cn}$ 1.5 3.0 6 10 15 20 25	I$_{cs}$ (1.5) (3.0) (6.0) (7.5) (7.5) (10.0) (12.5)

* Two rated short-circuit rating are defined in BS EN 60898 and BS EN 61009

(a)   I$_{cn}$ the rated short-circuit capacity (marked on the device).

(b)   I$_{cs}$ the service short-circuit capacity.

The difference between the two is the condition of the circuit-breaker after manufacturer's testing.

I$_{cn}$ is the maximum fault current the breaker can interrupt safely, although the breaker may no longer be usable.

I$_{cs}$ is the maximum fault current the breaker can interrupt safely without loss of performance.

The I$_{cn}$ value is normally marked on the device in a rectangle e.g. ⌷6000⌷ and for the majority of applications the prospective fault current at the terminals of the circuit-breaker should not exceed this value.

For domestic installations the prospective fault current is unlikely to exceed 6 kA up to which value the I$_{cn}$ and I$_{cs}$ values are the same.

The short-circuit capacity of devices to BS EN 60947-2 is as specified by the manufacturer.

(ii)   *circuit load characteristics*

(a)   semi-enclosed fuses - fuses should preferably be of the   533-01-04
cartridge type. Semi-enclosed fuses are still commonly
used in domestic and similar premises only

(b)   cartridge fuses to BS 1361 - these are for use in
domestic and similar premises

(c)   cartridge fuses to BS 88 - three types are specified:

gG   fuse links with a full-range breaking capacity for
general application

gM   fuse links with a full-range breaking capacity for
the protection of motor circuits

aM   fuse links with partial range breaking capacity for
the protection of motor circuits.

(d)   circuit-breakers to BS 3871-1 or BS EN 60898 and RCBOs
to BS EN 61009 - guidance on the selection is given in
Table 7.2B below.

TABLE 7.2B

cb type	Instantaneous trip current	Application
1 B	2.7 to 4 I_n 3 to 5 I_n	domestic and commercial installations having little or no switching surge
2 C 3	4.0 to 7.0 I_n 5 to 10 I_n 7 to 10 I_n	general use in commercial/industrial installations where the use of fluorescent lighting, small motors etc can produce switching surges that would operate a Type 1 or B circuit-breaker. Type C or 3 may be necessary in highly inductive circuits such as banks of fluorescent lighting
4 D	10 to 50 I_n 10 to 20 I_n	suitable for transformers X-ray machines, industrial welding equipment etc where high inrush currents may occur

Where I_n is the nominal rating of the device.

(iii)   *cable current-carrying capacities*

For guidance on the co-ordination of device and cable App 3 ratings see Appendix 6 App 4

(iv)   *disconnection times* 413-02-09
413-02-13

The protective device must operate within 0.4 or 5 seconds as appropriate for the circuit. Appendix 2 provides maximum permissible measured earth fault loop impedances for fuses, circuit-breakers and RCBOs.

7.3   Installation Considerations

7.3.1   Floors and ceilings

When a cable is installed under a floor or above a ceiling it shall be 522-06-05 run in such a position that it is not liable to damage by contact with the floor or ceiling or their fixings. Unarmoured cables passing through a joist shall be at least 50 mm from the top or bottom as appropriate or enclosed in earthed steel conduit. Alternatively, the cables can be provided with mechanical protection sufficient to prevent penetration of the cable by nails, screws and the like. (Note, the requirement to prevent penetration is difficult to meet.)

Fig 7.3.1: Cables through joists

cable in earthed
steel conduit

insulated
and
sheathed
cable

greater than
50 mm

joists

Notes:

1.   Maximum diameter of hole should be 0.25 x joist depth.

2.   Holes on centre line in a zone between 0.25 and 0.4 x span.

3.   Maximum depth of notch should be 0.125 x joist depth.

4.   Notches on top in a zone between 0.1 and 0.25 x span.

5.   Holes in the same joist should be at least 3 diameters apart.

7.3.2   Walls

Where a cable is concealed in a wall or partition at a depth of less 522-06-06 than 50 mm from any surface it must be enclosed in earthed metal conduit (trunking or ducting) or installed either horizontally within 150 mm of the top of the wall or partition or vertically within 150 mm of the angle formed by two walls, or run horizontally or vertically to an accessory or consumer unit (see Fig 7.3.2).

Fig 7.3.2: Permitted cable routes

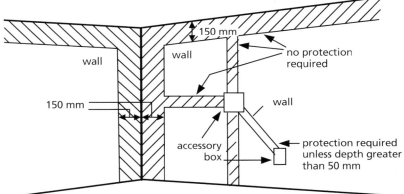

7.3.3   Telecommunication Circuits

528-01-02
528-01-04

An adequate separation between telecommunication wiring (Band I) and electric power and lighting (Band II) circuits must be maintained. This is to prevent mains voltage appearing in telecommunication circuits with consequent danger to personnel. BS 6701 : 1994 recommends that the minimum separation distances given in Tables 7.3A and 7.3B should be maintained:

TABLE 7.3A **External Cables**

Minimum separation distances between external low voltage electricity supply cables operating in excess of 50 V a.c. or 120 V d.c. to earth, but not exceeding 600 V a.c. or 900 V d.c. to earth (Band II), and Telecommunications cables (Band I).

Voltage to earth	Normal separation distances	Exceptions to normal separation distances, plus conditions to exception
Exceeding 50 V a.c. or 120 V d.c., but not exceeding 600 V a.c. or 900 V d.c.	50 mm	Below this figure a non-conducting divider should be inserted between the cables.

TABLE 7.3B **Internal Cables**

Minimum separation distances between internal low voltage electricity supply cables operating in excess of 50 V a.c. or 120 V d.c. to earth, but not exceeding 600 V a.c. or 900 V d.c. to earth (Band II) and Telecommunications cables (Band I).

Voltage to earth	Normal separation distances	Exceptions to normal separation distances, plus conditions to exception
Exceeding 50 V a.c. or 120 V d.c., but not exceeding 600 V a.c. or 900 V d.c.	50 mm	50 mm separation need not be maintained, provided that (i) the LV cables are enclosed in separate conduit which if metallic is earthed in accordance with BS 7671 **OR** (ii) the LV cables are enclosed in separate trunking which if metallic is earthed in accordance with BS 7671 **OR** (iii) the LV cable is of the mineral insulated type or is of earthed armoured construction

Notes:

1. Where the LV cables share the same tray then the normal separation should be met.

2. Where LV and telecommunications cables are obliged to cross additional insulation should be provided at the crossing point; this is not necessary if either cable is armoured.

7.3.4 Proximity to Other Systems
528-02

Electrical and all other services must be protected from any harmful mutual effects foreseen as likely under conditions of normal service. For example, cables should not be in contact 528-02-02 with or run alongside hot pipes.

The installation must comply with BS 7671, Chapter 52 and Chapter 54, regarding separation and bonding.

A particular form of harmful effect may occur when an electrical installation shares the space occupied by a hearing aid induction loop.

Under these circumstances, if phase(s) and neutral or switch feeds and switch wires are not close together, there may be interference with the induction loop.

This can occur when a conventional two-way circuit is installed. This effect can be reduced by connecting as shown in Fig 7.3.4.

Fig 7.3.4: Circuit for reducing interference with induction loop

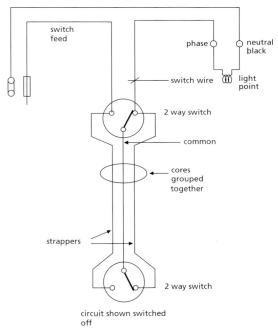

7.3.5 Height of switches, sockets etc

Accessories for general use, such as light switches and socket-outlets, are required by the Building Regulations to be located so that they can be used by people whose reach is limited. A way of satisfying this requirement is to install switches for lights and similar equipment and socket-outlets 553-01-06 at a height of between 450 mm and 1200 mm from finished floor level. See Figure 8A in Appendix 8.

55

7.4   Smoke Alarms and Emergency Lighting

7.4.1   Smoke alarms in single dwellings

The Building Regulations 1991 and the Building Standards (Scotland) Regulations 1990 require all new and refurbished dwellings to be fitted with mains operated smoke alarms. The requirements for single family dwellings of not more than two storeys are that self-contained smoke alarms should be installed as follows:

1)   at least one on each floor

2)   within 7 m of kitchens and living rooms or other areas where fires may start e.g. integral garages

3)   within 3 m of all bedroom doors.

The smoke alarm must be installed in accordance with the manufacturer's instructions, generally on ceilings and at least 300 mm from any wall or ceiling luminaire.

The alarms are required to:-

-    be interconnected so that detection of smoke by one unit operates the alarm in all units

-    be permanently wired to a separate way in the distribution board (consumer unit), or supplied from a local, regularly used lighting circuit.

-    have battery backup

Except for TT supplies, the circuit supplying a smoke alarm should not be protected by an RCD. For TT supplies the smoke alarm must be supplied from the fixed equipment section, that is <u>not</u> through a 30 mA RCD, see Figure 3a.

The cable for the power supply to each self-contained unit and for the interconnections need have no fire retardant properties, and needs no special segregation. Otherwise, fire 528-01-04 alarm system cables generally are required to be segregated as per BS 5839-1 and BS 5839-6, see Table 7.4.

TABLE 7.4 Segregation requirements of fire alarm and emergency lighting standards

Fire alarms BS 5839		Emergency lighting BS 5266	
a	installation in conduit, ducting, trunking or a channel reserved for fire alarms, or	a	physical segregation by a minimum distance of 300 mm, or
b	a mechanically strong, rigid and continuous partition of non-combustible material, or	b	use of mineral cables, or
c	mounting at a distance of at least 300 mm from conductors of other systems, or	c	cables to BS 6387 Cat B
d	wiring in cables complying with BS 7629, or		
e	wiring in mineral insulated copper sheathed cable with an insulating sheath or barrier. The exposed-to-touch rating of the IEE Wiring Regulations should not be exceeded.		Table 4J1A

7.4.2 Emergency lighting

The wiring to self-contained emergency lighting luminaires is not considered part of the emergency lighting installation and needs no special segregation. Otherwise, emergency lighting 528-01-04 and fire alarm circuits must be separated from other cables and from each other, in compliance with BS 5266, see Table 7.4.

SECTION 8. SPECIAL LOCATIONS GIVING RISE TO INCREASED RISK OF ELECTRIC SHOCK

8.1 Locations Containing a Bath or Shower

8.1.1   Because of the presence of water these locations are onerous for equipment and there is an increased danger of electric shock because of immersion of the body in water.

The additional requirements can be summarised as follows:

(i)   No socket-outlets are allowed other than SELV and shaver supply units – see Table 8.1.   601-08-01

(ii)   Supplementary bonding of the terminals of the protective conductors of circuits to Class I and Class II equipment in the zones (see Figure 4d) to exposed-conductive-parts in the zones is required, including: 601-04-01 601-04-02
- metal pipes both water and central heating
- metal baths and shower basins
- accessible metal structural parts of the building.
The supplementary bonding must be carried out in or in close proximity to the zones. See Figures 4d and 4e.

(iii)   Protection against ingress of moisture is specified for equipment installed in the zones – see figure 4d and Table 8.1. The requirements apply to appliances, switchgear and wiring accessories.   601-06-01

(iv)   There are restrictions as to where appliances, switchgear and wiring accessories may be installed – see Table 8.1.   601-08 601-09

8.1.2   Underfloor heating installations in these areas should have an overall earthed metallic grid or the heating cable should have an earthed metallic sheath, which must be supplementary bonded.   601-09-04

TABLE 8.1
Requirements for equipment (current using and accessories) in locations containing a bath or shower

Zone note 2	Minimum degree of protection	Requirements for equipment in the zones		
		Current Using (Appliances)	Switchgear and Accessories	
0	IPX7	Only SELV fixed equipment that cannot be located elsewhere.	None allowed.	601-08-01 601-09-01
1	IPX4	SELV equipment allowed. Water heaters, showers, shower pumps, allowed. Other fixed equipment that cannot reasonably be located elsewhere allowed if protected by a 30 mA RCD.	Only 12 V a.c. and 30 V d.c. switches of SELV circuits allowed, the source being outside zones 0, 1 and 2.	601-08-01 601-09-01
2	IPX4	SELV equipment allowed. Water heaters, showers, shower pumps, luminaires, fans, heating appliances, units for whirlpool baths allowed. Other fixed equipment that cannot reasonably be located elsewhere allowed.	SELV switches and sockets allowed, the source being outside zones 0, 1 and 2, and shaver supply units to BS EN 60742 Ch 2 Sec 1 allowed only if fixed where direct spray from showers is unlikely.	601-08-01 601-09-03
3	No requirement.	SELV equipment allowed. Appliances allowed and, unless fixed, must be protected by a 30 mA RCD.	Accessories allowed except for socket-outlets. There is to be no provision for connecting portable equipment. SELV sockets and shaver supply units to BS EN 60742 Chap 2 Sect 1 allowed.	601-08-01 601-09-03
Outside Zones	No requirement.	Appliances allowed	Accessories allowed except for socket-outlets. SELV sockets and shaver supply units to BS EN 60742 Chap 2 Sect 1 allowed.	601-08-01
				601-08-02 412-06

Note 1: Where a shower cubicle is installed in a room other than a bathroom or shower room, outside zones 0, 1, 2 or 3 a socket-outlet, other than a SELV socket-outlet or shaver supply unit, shall be protected by a residual current device with rated residual operating current ($I_{\Delta n}$) not exceeding 30 mA in accordance with Regulation 412-06.

Note 2: See Figures 4d and 4e for zones.

8.2 Shower Cubicles in a Room used for Other Purposes

Where a shower cubicle is installed in a room other than a bathroom or shower room the requirements for bathrooms and shower rooms are generally to be complied with, except that socket-outlets are allowed outside zones 0, 1, 2 and 3 provided they are protected by a 30 mA RCD, and no supplementary bonding is required in zone 3.

601-04-02

8.3 Temporary and Garden Buildings, Domestic Garages, Buildings of Lightweight Construction etc

300-01

The use of a temporary building does not permit a lower standard of electrical installation. The standards of installation and maintenance need to be higher to cope with the more onerous conditions. Particular attention must be paid to:

(i)   suitability of the equipment for the environment
(ii)   earthing and bonding
(iii)   connection to the supply
(iv)   use of accessories of the appropriate Degree of Protection (IP code) to suit the particular external influences.

8.4 Earthing requirements of Equipment having High Protective Conductor Current

607

Equipment

Equipment having a protective conductor current exceeding 3.5 mA shall be either permanently connected to the fixed wiring or connected by means of an industrial plug and socket to BS EN 60309-2.

607-02-02

Equipment having a protective conductor current exceeding 10 mA shall be connected preferably by a permanent connection, or an industrial plug and socket to BS EN 60309-2 with a protective conductor csa of at least 2.5 mm^2 for plugs up to 16 A and at least 4 mm^2 for plugs rated above to 16 A.

607-02-03

Circuits

The wiring of every final circuit and distribution circuit having a protective conductor current likely to exceed 10 mA shall have high integrity protective conductor connections comprising either:

607-02-04

(i)   a single copper protective conductor complying with Section 543 and of csa not less than 10 mm^2, or

(ii)   a single copper protective conductor complying with Section 543 and of csa not less than 4 mm² enclosed in conduit, or

(iii)   duplicate protective conductors, each complying with Section 543.

Socket-outlet final circuits

For socket-outlet final circuits requiring a high integrity protective 607-03-01 conductor connection, the following arrangements are acceptable:

(i)   a ring circuit with a ring protective conductor (Fig 8.4a), or

(ii)   a radial circuit with a single protective conductor (Fig 8.4b) connected as a ring or an additional protective conductor provided by conduit, trunking or ducting.

When the two protective conductors are provided the ends must be terminated independently of each other at all connection points e.g. distribution board and socket-outlet. Accessories are required to have two separate earth terminals.

Fig 8.4a: Ring final circuit supplying socket-outlets
(total protective conductor current exceeding 10 mA)

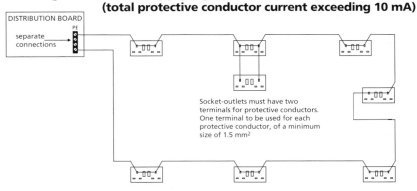

Fig 8.4b: Radial circuit supplying socket-outlets
with duplicate protective conductor
(total protective conductor current exceeding 10 mA)

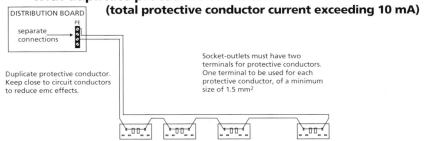

SECTION 9.   INSPECTION AND TESTING

Part 7

9.1   Inspection and Testing

Every installation must be inspected and tested during erection 711-01-01
and on completion before being put into service.

Precautions shall be taken to avoid danger to persons and to 711-01-01
avoid damage to property and installed equipment during
inspection and testing.

If the inspection and tests are satisfactory, a signed Electrical 741
Installation Certificate together with a Schedule of Inspections
and a Schedule of Test Results (as in Appendix 7) are to be
given to the person ordering the work.

9.2   Inspection

9.2.1   Procedure and Purpose

Inspection shall precede testing and shall normally be done 712-01-01
with that part of the installation under inspection disconnected
from the supply.

The purpose of the inspection is to verify that equipment is: 712-01-02

(i)   correctly selected and erected in accordance with
         BS 7671 (and if appropriate its own standard)

(ii)   not visibly damaged or defective so as to impair safety.

9.2.2   Inspection Checklist   712-01-03

The inspection shall include at least the checking of relevant
items from the following checklist:

(i)   connection of conductors

(ii)   identification of conductors

(iii)   routing of cables in safe zones or protected against
         mechanical damage

(iv)   selection of conductors for current-carrying capacity
         and voltage drop, in accordance with the design

(v)   connection of single-pole devices for protection or
         switching in phase conductors only

(vi)   correct connection of accessories and equipment (including polarity)

(vii)   presence of fire barriers, suitable seals and protection against thermal effects

(viii)   methods of protection against electric shock

    (a)   protection against both direct contact and indirect contact, i.e.:

        - SELV
        - limitation of discharge of energy

    (b)   protection against direct contact (including measurement of distances where appropriate), i.e.:

        - protection by insulation of live parts
        - protection by barriers or enclosure
        - protection by PELV

    (c)   protection against indirect contact:

        - earthed equipotential bonding and automatic disconnection of supply
            presence of earthing conductor
            presence of protective conductors
            presence of main equipotential bonding conductors
            presence of supplementary equipotential bonding conductors
        - use of Class II equipment or equivalent insulation
        - electrical separation

(ix)   prevention of mutual detrimental influence   515

Account must be taken of the proximity of other 528 electrical services of a different voltage band and of non-electrical services and influences.

Fire alarm and emergency lighting circuits must be 528-01-04 separated from other cables and from each other, in compliance with BS 5839 and BS 5266, see 7.4.1 and 7.4.2.

Band I and Band II circuit cables may not be present in 528-01-02
the same enclosure or wiring system unless they are
either separated by an effective barrier or wired with
cables suited to the highest voltage present. Where 528-01-07
common boxes are used for Band I and Band II circuits,
the circuits must be segregated by an insulating or
earthed partition.

Mixed categories of circuits may be contained in 528-01-02
multicore cables, subject to specific requirements.

Definitions of voltage bands   Part 2

Band I circuit:

Circuits that are nominally extra-low i.e. not exceeding 50 V
ac or 120 V dc e.g. SELV, PELV, telecommunications, data
and signalling.

Band II circuit:

Circuits that are nominally low voltage, that is 51 to
1000 V ac and 120 to 1500 V dc. Telecommunication
cables that are generally ELV but have ringing voltages
exceeding 50 V are Band I.

(x)   presence of appropriate devices for isolation and   131-14-01
        switching correctly located

(xi)   presence of undervoltage protective devices (where   451
        appropriate)

(xii)   choice and setting of protective and monitoring devices
        (for protection against indirect contact and/or
        protection against overcurrent)

(xiii) labelling of circuits, cbs, RCDs, fuses, switches and   514
        terminals, main earthing and bonding connections

(xiv)   selection of equipment and protective measures   522
        appropriate to external influences

(xv)   adequacy of access to switchgear and equipment

(xvi)   presence of danger notices and other warning signs
        (see Section 6)

(xvii)	presence of diagrams, instructions and similar information	514-09
(xviii)	erection methods	522
(xix)	requirements for special locations.	600

9.3 Testing

713

Testing must include the relevant tests from the following checklist.

When a test shows a failure to comply, the installation must be corrected. The test must then be repeated, as must any earlier test that could have been influenced by the failure.

713-01-01

9.3.1 Testing Checklist

(i) continuity of protective conductors (including main and supplementary equipotential bonding conductors)

713-02-01

(ii) continuity of ring final circuit conductors including protective conductors

713-03-01

(iii) insulation resistance (between live conductors and between each live conductor and earth)

713-04

(iv) polarity; this includes checks that single-pole control and protective devices (e.g. switches, circuit-breakers, fuses) are connected in the phase conductor only, that bayonet and Edison-screw lampholders (except for E14 and E27 to BS EN 60238) have their outer contacts connected to the neutral conductor and that wiring has been correctly connected to socket-outlets and other accessories

713-09

(v) earth electrode resistance

713-10

(vi) earth fault loop impedance

713-11

(vii) prospective fault current, if not determined by enquiry of the electricity supplier

713-12

(viii) functional testing (including RCDs and RCBOs).

713-13

SECTION 10. GUIDANCE NOTES ON INITIAL TESTING OF INSTALLATIONS

10.1 Safety and equipment 711-01-01

Electrical testing involves danger. It is the tester's duty to ensure his or her own safety, and the safety of others, in the performance of the test procedures. When using test instruments, this is best achieved by precautions such as:

(i)   an understanding of the correct application and use of the test instrumentation, leads, probes and accessories to be employed

(ii)   checking that the test instrumentation is made in accordance with the appropriate safety standards such as BS EN 61243-3 for two-pole voltage detectors and BS EN 61010 or BS EN 61557 for instruments

(iii)   checking before use that all leads, probes, accessories (including all devices such as crocodile clips used to attach conductors) and instruments are clean, undamaged and functioning

(iv)   observing the safety measures and procedures set out in HSE Guidance Note GS 38 for all instruments, leads, probes and accessories. It should be noted that some test instrument manufacturers advise that their instruments be used in conjunction with fused test leads and probes. Other manufacturers advise the use of non-fused leads and probes when the instrument has in-built electrical protection, but it should be noted that such electrical protection does not extend to the probes and leads.

10.2 Sequence of Tests 713-01-01

Note: The advice given does not preclude other test methods.

Tests should be carried out in the following sequence:

10.2.1 Before the supply is connected

(i)   continuity of protective conductors, including main and supplementary bonding

(ii)   continuity of ring final circuit conductors, including protective conductors

(iii)   insulation resistance

(iv)   polarity (by continuity methods)

(v)   earth electrode resistance, when using an earth electrode resistance tester (see also vii).

10.2.2 With the supply connected

(vi)   re-check of polarity

(vii)   earth electrode resistance, when using a loop impedance tester

(viii)   earth fault loop impedance

(ix)   prospective fault current measurement, if not determined by enquiry of the electricity supplier

(x)   functional testing.

Results obtained during various tests should be recorded in the 741-01-01 Schedule of Test Results (Appendix 7) for future reference.

10.3   Test Procedures

10.3.1 Continuity of protective and bonding conductors   713-02-01 (except ring final circuits, see Para 10.3.2)

Test Methods 1 and 2 are alternative ways of testing the continuity of protective conductors.

Every protective conductor including the earthing conductor, main and supplementary bonding conductors should be tested to verify that the conductors are electrically sound and correctly connected.

Test Method 1 detailed below, as well as checking the continuity of the protective conductor, also measures $(R_1 + R_2)$ which, when added to the external impedance (Z_e), enables the earth-fault loop impedance (Z_s) to be checked against the design, see Section 10.3.6. Note: $(R_1 + R_2)$ is the sum of the resistances of the phase conductor (R_1) and the circuit protective conductor (R_2) between the point of utilisation and origin of the installation.

Use an ohmmeter capable of measuring a low resistance for these tests.

Test Method 1 can only be used to measure $(R_1 + R_2)$ for an 'all insulated' installation. Installations incorporating steel conduit, steel trunking, micc and pvc/swa cables will produce parallel paths to protective conductors. Such installations should be inspected for soundness of construction and Test Method 1 or 2 used to prove continuity.

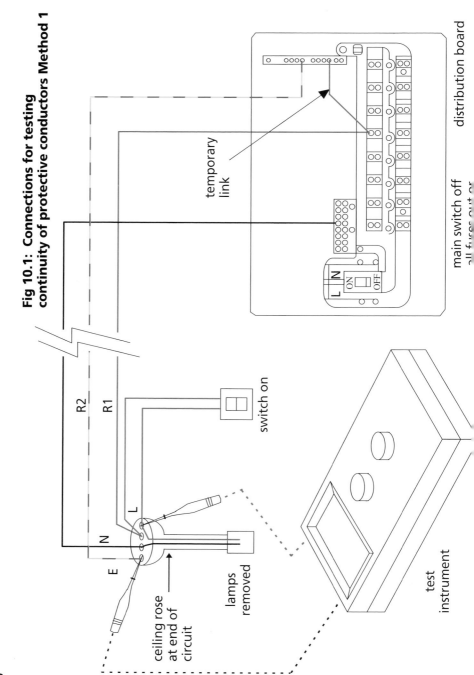

Fig 10.1: Connections for testing continuity of protective conductors Method 1

temporary link

distribution board

main switch off
all fuses out or

R2

R1

switch on

L

N

E

ceiling rose
at end of
circuit

lamps
removed

test
instrument

10.3.1(i) To test the continuity of protective conductors 713-02-01

Test Method 1

Bridge the phase conductor to the protective conductor at the distribution board so as to include all the circuit. Then test between phase and earth terminals at each point in the circuit. The measurement at the circuit's extremity should be recorded and is the value of $(R_1 + R_2)$ for the circuit under test (see Fig 10.1).

If the instrument does not include an "auto-null" facility, or this is not used, the resistance of the test leads should be measured and deducted from the resistance readings obtained.

Test Method 2

Connect one terminal of the continuity test instrument to a long test lead and connect this to the consumer's main earthing terminal.

Connect the other terminal of the instrument to another test lead and use this to make contact with the protective conductor at various points on the circuit, such as luminaires, switches, spur outlets etc.

The resistance of the protective conductor R_2 is recorded on the Schedule of Test Results, form F4.

10.3.1(ii) To test the continuity of bonding conductors 713-02-01

Use Test Method 2

10.3.2 Continuity of ring final circuit conductors 713-03-01

A three step test is required to verify the continuity of the phase, neutral and protective conductors and correct wiring of every ring final circuit. The test results show if the ring has been inter-connected to create an apparently continuous ring circuit which is in fact broken, or wrongly wired.

Step 1:
The phase, neutral and protective conductors are identified and the end-to-end resistance of each is measured separately (see Fig10.2a). These resistances are r_1, r_n and r_2 respectively. A finite reading confirms that there is no open circuit on the ring conductors under test. The resistance values obtained should be the same (within 0.05 ohm) if the conductors are the same

size. If the protective conductor has a reduced csa the resistance r_2 of the protective conductor loop will be proportionally higher than that of the phase and neutral loops e.g. 1.67 times for 2.5/1.5 mm^2 cable. If these relationships are not achieved then either the conductors are incorrectly identified or there is something wrong at one or more of the accessories.

Step 2:
The phase and neutral conductors are then connected together so that the outgoing phase conductor is connected to the returning neutral conductor and vice-versa (see Fig 10.2b). The resistance between phase and neutral conductors is measured at each socket-outlet. The readings at each of the sockets wired into the ring will be substantially the same and the value will be approximately one quarter of the resistance of the phase plus the neutral loop resistances, i.e. $(r_1 + r_n)/4$. Any sockets wired as spurs will have a higher resistance value due to the resistance of the spur conductors.

Note: Where single-core cables are used, care should be taken to verify that the phase and neutral conductors of opposite ends of the ring circuit are connected together. An error in this respect will be apparent from the readings taken at the socket-outlets, progressively increasing in value as readings are taken towards the midpoint of the ring, then decreasing again towards the other end of the ring.

Step 3:
The above step is then repeated, this time with the phase and cpc cross-connected (see Fig 10.2c). The resistance between phase and earth is measured at each socket. The readings obtained at each of the sockets wired into the ring will be substantially the same and the value will be approximately one quarter of the resistance of the phase plus cpc loop resistances, i.e. $(r_1 + r_2)/4$. As before, a higher resistance value will be recorded at any sockets wired as spurs. The highest value recorded represents the maximum $(R_1 + R_2)$ of the circuit and is recorded on Form F4. The value can be used to determine the earth loop impedance (Zs) of the circuit to verify compliance with the loop impedance requirements of BS 7671 (see Appendix 9).

This sequence of tests also verifies the polarity of each socket, except that if the testing has been carried out at the terminals on the reverse of the accessories, a visual inspection is required to confirm correct polarity connections, and dispenses with the need for a separate polarity test.

Fig 10.2: Connections for testing continuity of ring final circuit conductors

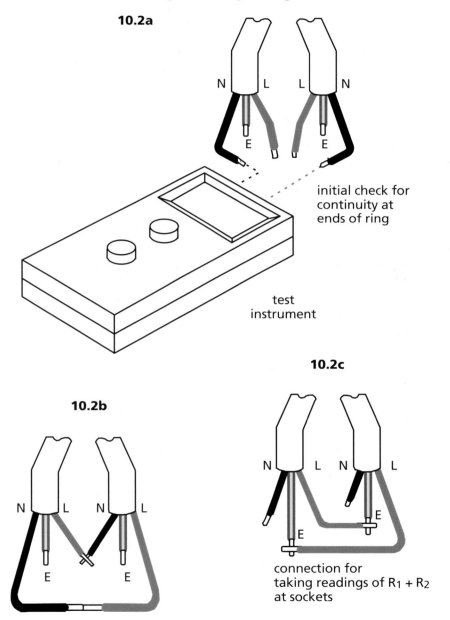

10.2a

N L L N

E E

initial check for
continuity at
ends of ring

test
instrument

10.2b

N L N L

E E

10.2c

N L N L

E

E

connection for
taking readings of $R_1 + R_2$
at sockets

Pre-test checks

10.3.3(i)

(a) pilot or indicator lamps, and capacitors are disconnected from circuits to avoid misleading test values being obtained

(b) if a circuit includes voltage-sensitive electronic devices such as dimmer switches, touch switches, delay timers, power controllers, electronic starters or controlgear for fluorescent lamps etc either:

(1) the devices must be temporarily disconnected, or
(2) a measurement should be made between live conductors (phase and neutral) connected together and the protective earth only

It should also be confirmed that there are no RCDs incorporating electronic amplifiers, before the test is made.

10.3.3(ii)

Tests should be carried out using the appropriate d.c. test voltage specified in Table 10.1.

The tests should be made at each distribution with the main switch off, all fuses in place, switches and circuit-breakers closed, lamps removed and other current-using equipment disconnected. Where the removal of lamps and/or the disconnection of current-using equipment is impracticable, the local switches controlling such lamps and/or equipment should be open.

Where any circuits contain two-way switching the two-way switches must be operated one at a time and further insulation resistance tests carried out to ensure that all the circuit wiring is tested.

TABLE 10.1 TABLE 71A

Minimum values of insulation resistance

Circuit nominal voltage	Test voltage V d.c.	Minimum insulation resistance (M ohms)
SELV and PELV	250	0.25
Up to and including 500 V with the exception of SELV and PELV, but including FELV	500	0.5

Although an insulation resistance value of not less than 0.5 megohm complies with BS 7671, where an insulation resistance of less than 2 megohms is recorded the possibility of a latent defect exists. Each circuit should then be tested separately, and its insulation resistance should be greater than 2 megohms.

Where electronic devices are disconnected for the purpose of 713-04-04 the tests on the installation wiring (and the devices have exposed-conductive-parts required by BS 7671 to be connected to the protective conductors) the insulation resistance between the exposed-conductive-parts and all live parts of the device (phase and neutral connected together) should be measured separately and should not be less than the values stated in Table 10.1.

10.3.3(iii) *Insulation resistance between live conductors* 713-04

Single-phase and three-phase

Test between all the live (phase and neutral) conductors at the distribution board (see Fig 10.3).

Fig 10.3: Insulation resistance tests between live conductors of a circuit

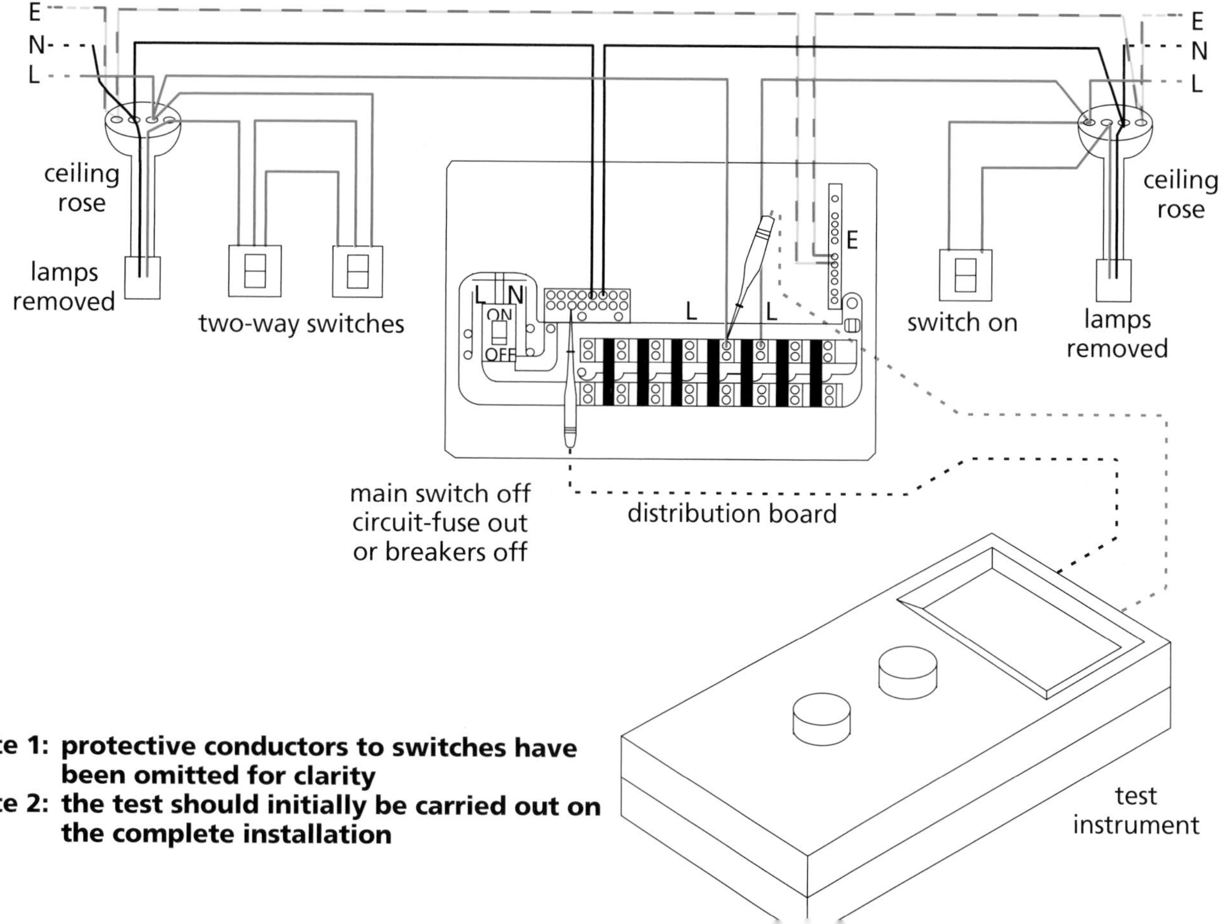

ceiling rose

lamps removed

switch on

two-way switches

ceiling rose

lamps removed

distribution board

main switch off
circuit-fuse out
or breakers off

test instrument

ON

OFF

E

N

L

E

N

L

L

L

E

**note 1: protective conductors to switches have
been omitted for clarity**
**note 2: the test should initially be carried out on
the complete installation**

74

Resistance readings obtained should be not less than the minimum values referred to in Table 10.1.

10.3.3(iv) *Insulation resistance to Earth* 713-04

Single-phase

Test between the live conductors (phase and neutral) and the circuit protective conductors at the distribution board (see Fig 10.4).

For circuits containing two-way switching or two-way and intermediate switching the switches must be operated one at a time and the circuit subjected to additional insulation resistance tests.

Three-phase

Test to earth from all live conductors (including the neutral) connected together. Where a low reading is obtained it is necessary to test each conductor separately to earth, after disconnecting all equipment.

Resistance readings obtained should be not less than the minimum values referred to in Table 10.1.

10.3.3(v) *SELV and PELV circuits* Table 71A

Test between SELV and PELV circuits and live parts of other circuits at 500 V d.c.

Test between SELV or PELV conductors at 250 V d.c. and between PELV conductors and protective conductors of the PELV circuit at 250 V d.c.

10.3.3(vi) *FELV circuits* 471-14-03

FELV circuits are tested as LV circuits at 500 V d.c.

Fig 10.4: Insulation resistance tests to earth

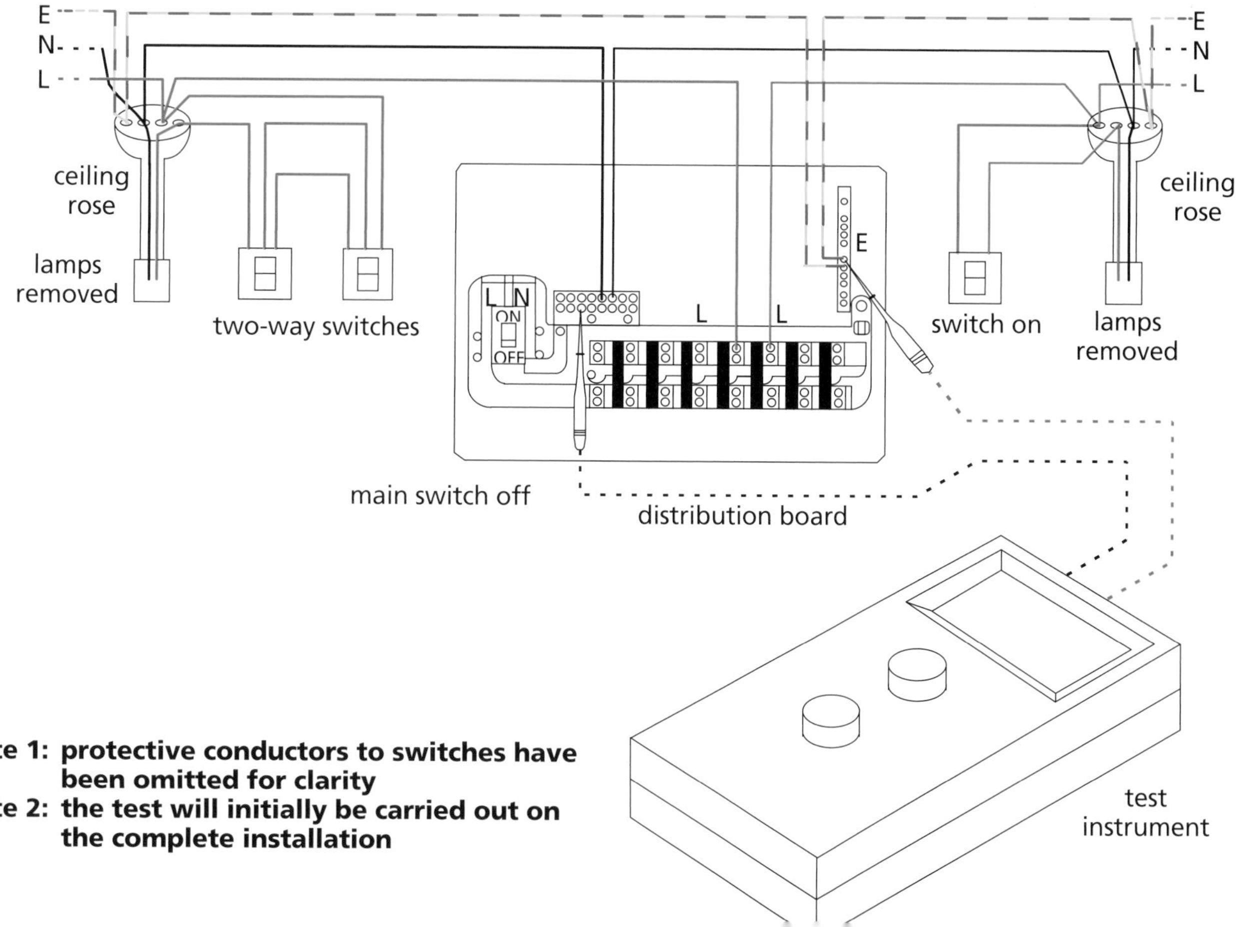

ceiling rose

E — N — L

lamps removed

switch on

E

L L

distribution board

main switch off

L N
ON
OFF

two-way switches

ceiling rose

lamps removed

E — N — L

test instrument

note 1: protective conductors to switches have been omitted for clarity

note 2: the test will initially be carried out on the complete installation

Fig 10.5: Polarity test on a lighting circuit

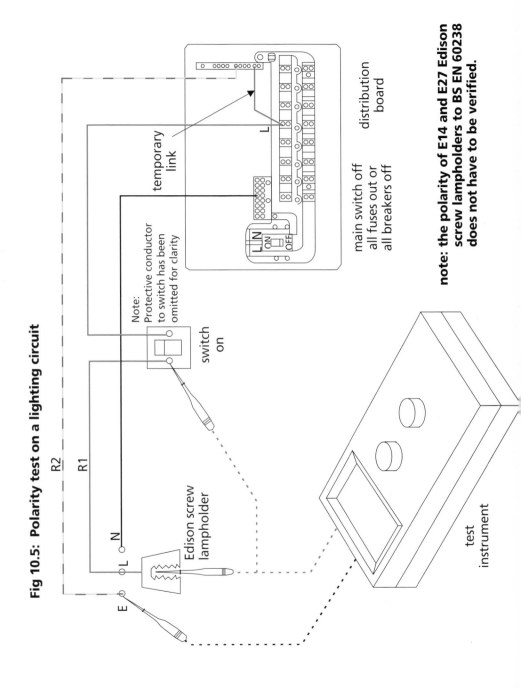

temporary link

distribution board

main switch off
all fuses out or
all breakers off

Note:
Protective conductor
to switch has been
omitted for clarity

switch on

R2

R1

N

L

E

Edison screw lampholder

test instrument

note: the polarity of E14 and E27 Edison screw lampholders to BS EN 60238 does not have to be verified.

10.3.4 Polarity 713-09

See Figure 10.5.

The method of test prior to connecting the supply is the same as Test Method 1 for checking the continuity of protective conductors which should have already been carried out (see 10.3.1, 10.3.2 and Figs 10.1 and 10.2). For ring circuits a visual check may be required (see 10.3.2 following step 3).

It is important to confirm that:

(i)   overcurrent devices and single-pole controls are in the phase conductor 713-09-01

(ii)   except for E14 and E27 lampholders to BS EN 60238, centre contact screw lampholders have the outer threaded contact connected to the neutral and 713-09-01

(iii) socket polarities are correct. 713-09-01

After connection of the supply polarity must be checked using a voltmeter or a test lamp (both with leads complying with HSE Guidance Note GS 38).

10.3.5 Earth Electrode Resistance 713-10

If the electrode under test is being used in conjunction with an RCD protecting an installation, the following method of test may be applied.

A loop impedance tester is connected between the phase conductor at the origin of the installation and the earth electrode <u>with the test link open</u>, and a test performed. This impedance reading is treated as the electrode resistance and is then added to the resistance of the protective conductor for the protected circuits. The test should be carried out before energising the remainder of the installation.

The measured resistance should meet the following criteria and those of 10.3.6 but in any case should not exceed 200 ohms: 542-02-02

For TT systems, the value of the earth electrode resistance R_A in ohms multiplied by the operating current in amperes of the protective device $I_{\Delta n}$ shall not exceed 50 V e.g. if $R_A = 200\ \Omega$, then the maximum RCD operating current should not exceed 250 mA. 413-02-20

Remember to replace the test link.

The earth fault loop impedance (Z_s) is required to be 413-02-10
determined for the furthest point of each circuit. It may be 413-02-11
determined 413-02-14

- by direct measurement of Z_s

- by direct measurement of Z_e at the origin and adding ($R_1 + R_2$) measured during the continuity tests (10.3.1 and 10.3.2) ($Z_s = Z_e + (R_1 + R_2)$)

- by adding ($R_1 + R_2$) measured during the continuity tests to the value of Z_e declared by the electricity supplier, (see 7.1(i)). The effectiveness of the supplier's earth must be confirmed by a test.

The external impedance (Z_e) may be measured using a phase-earth loop impedance tester.

The main switch is opened and made secure to disconnect the installation from the source of supply. The earthing conductor is disconnected from the main earthing terminal and the measurement made between phase and earth of the supply.

Remember to reconnect the earthing conductor to the earth terminal after the tests.

Direct measurement of Z_s can only be made on a live installation. Neither the connection with earth nor bonding conductors are disconnected. The reading given by the loop impedance tester will usually be less than $Z_e + (R_1 + R_2)$ because of parallel earth return paths provided by any bonded extraneous-conductive-parts. This must be taken into account when comparing the results with design data.

Care should be taken to avoid any shock hazard to the testing personnel and to other persons on site during the tests.

The values of Z_s determined should be less than the value given 413-02-08
in Appendix 2 for the particular overcurrent device and cable. 543-01-03

For TN systems, when protection is afforded by an rcd, the 413-02-16
rated residual operating current in amperes times the earth fault loop impedance in ohms should not exceed 50 V. This test should be carried out before energising other parts of the system.

Note: For further information on the measurement of earth fault loop impedance, refer to Guidance Note No 3 — Inspection and Testing.

10.3.7 Measurement of prospective fault current 712-12-01

It is not recommended that installation designs are based on 434-02-01
measured values of prospective fault current, as changes to the
supply network subsequent to the completion of the
installation may increase fault levels.

Designs should be based on the maximum fault current 313-01-01
provided by the electricity supplier (see 7.2.5(i)).

If it is desired to measure prospective fault levels this should be
done with all main bonding in place. Measurements are made
at the distribution board between live conductors and between
phase conductors and earth.

For three-phase supplies the maximum possible fault level will
be approximately twice the single-phase to neutral value. (For
three-phase to earth faults, neutral and earth path impedances
have no influence.)

10.3.8 Functional testing 713-13

RCDs should be tested as described in Section 11. All assemblies 713-13-02
including switchgear, controls, and interlocks should be
functionally tested; that is, operated to check that they work
and are properly fixed etc.

SECTION 11. OPERATION OF RESIDUAL CURRENT OPERATED DEVICES (RCDs) AND RESIDUAL CURRENT BREAKERS WITH OVERCURRENT PROTECTION (RCBOs)

713-13-01

11.1 General Test Procedure

The tests are made on the load side of the RCD, as near as practicable to its point of installation, and between the phase conductor of the protected circuit and the associated circuit protective conductor. The load supplied should be disconnected during the test.

11.2 General purpose RCDs to BS 4293

(i) with a leakage current flowing equivalent to 50 % of the rated tripping current, the device should not open.

(ii) with a leakage current flowing equivalent to 100 % of the rated tripping current of the RCD, the device should open in less than 200 ms. Where the RCD incorporates an intentional time delay it should trip within a time range from '50 % of the rated time delay plus 200 ms' to '100 % of the rated time delay plus 200 ms'.

11.3 General purpose RCCBs to BS EN 61008 or RCBOs to BS EN 61009

(i) with a leakage current flowing equivalent to 50 % of the rated tripping current of the RCD the device should not open.

(ii) with a leakage current flowing equivalent to 100 % of the rated tripping current of the RCD, the device should open in less than 300 ms unless it is of 'Type S' (or selective) which incorporates an intentional time delay. In this case, it should trip within a time range from 130 ms to 500 ms

11.4 RCD protected socket-outlets to BS 7288

(i) with a leakage current flowing equivalent to 50 % of the rated tripping current of the RCD the device should not open

(ii) with a leakage current flowing equivalent to 100 % of the rated tripping current of the RCD, the device should open in less than 200 ms

11.5   Additional Requirement for Supplementary Protection 412-06-02

Where an RCD or RCBO with a rated residual operating current $I_{\Delta n}$ not exceeding 30 mA is used to provide supplementary protection against direct contact, with a test current of $5I_{\Delta n}$ the device should open in less than 40 ms. The maximum test time must not be longer than 40 ms, unless the protective conductor potential rises by less than 50 V. (The instrument supplier will advise on compliance).

11.6   Integral Test Device 713-13-02

An integral test device is incorporated in each RCD. This device enables the electrical and mechanical parts of the RCD to be verified, by pressing the button marked 'T' or 'Test'.

Operation of the integral test device does not provide a means of checking:

(a)   the continuity of the earthing conductor or the associated circuit protective conductors, or

(b)   any earth electrode or other means of earthing, or

(c)   any other part of the associated installation earthing.

The test button will only operate the RCD if the RCD is energised.

Confirm that the notice to test RCDs quarterly (by pressing the 514-12-02 test button) is fixed in a prominent position ((see 6.1(xi)).

APPENDICES
CONTENTS

MAXIMUM DEMAND AND DIVERSITY

This Appendix gives some information on the determination of the maximum demand for an installation and includes the current demand to be assumed for commonly used equipment. It also includes some notes on the application of allowances for diversity.

The information and values given in this Appendix are intended only for guidance because it is impossible to specify the appropriate allowances for diversity for every type of installation and such allowances call for special knowledge and experience. The figures given in Table 1B, therefore, may be increased or decreased as decided by the engineer responsible for the design of the installation concerned. For blocks of residential dwellings, large hotels, industrial and large commercial premises, the allowances are to be assessed by a competent person.

The current demand of a final circuit is determined by summating the current demands of all points of utilisation and equipment in the circuit and, where appropriate, making an allowance for diversity. Typical current demands to be used for this summation are given in Table 1A.

The current demand of a circuit supplying a number of final circuits may be assessed by using the allowances for diversity given in Table 1B which are applied to the total current demand of all the equipment supplied by that circuit and not by summating the current demands of the individual final circuits obtained as outlined above. In Table 1B the allowances are expressed either as percentages of the current demand or, where followed by the letters f.l., as percentages of the rated full load current of the current-using equipment. The current demand for any final circuit which is a conventional circuit arrangement complying with Appendix 8 is the rated current of the overcurrent protective device of that circuit.

An alternative method of assessing the current demand of a circuit supplying a number of final circuits is to summate the diversified current demands of the individual circuits and then apply a further allowance for diversity. In this method the allowances given in Table 1B are not to be used, the values to be chosen being the responsibility of the designer of the installation.

The use of other methods of determining maximum demand is not precluded where specified by a suitably qualified electrical engineer. After the design currents for all the circuits have been determined, enabling the conductor sizes to be chosen, it is necessary to check that the limitation on voltage drop is met.

TABLE 1A
Current demand to be assumed for points of utilisation and current-using equipment

Point of utilisation or current-using equipment	Current demand to be assumed
Socket-outlets other than 2 A socket-outlets and other than 13 A socket-outlets see note 1	Rated current
2 A socket-outlets	At least 0.5 A
Lighting outlet see note 2	Current equivalent to the connected load, with a minimum of 100 W per lampholder
Electric clock, shaver supply unit (complying with BS 3535), shaver socket-outlet (complying with BS 4573), bell transformer, and current-using equipment of a rating not greater than 5 VA	May be neglected
Household cooking appliance	The first 10 A of the rated current plus 30 % of the remainder of the rated current plus 5 A if a socket-outlet is incorporated in the control unit
All other stationary equipment	British Standard rated current, or normal current

Note 1: See Appendix 8 for the design of standard circuits using socket-outlets to BS 1363-2 and BS 4343.

Note 2: Final circuits for discharge lighting must be arranged so as to be capable of carrying the total steady current, viz. that of the lamp(s) and any associated gear and also their harmonic currents. Where more exact information is not available, the demand in volt-amperes is taken as the rated lamp watts multiplied by not less than 1.8. This multiplier is based upon the assumption that the circuit is corrected to a power factor of not less than 0.85 lagging, and takes into account control gear losses and harmonic current.

TABLE 1B
Allowances for diversity

Purpose of final circuit fed from conductors or switchgear to which diversity applies	Type of premises		
	Individual household installations including individual dwellings of a block	Small shops, stores, offices and business premises	Small hotels, boarding houses, guest houses, etc
1. Lighting	66 % of total current demand	90 % of total current demand	75 % of total current demand
2. Heating and power (but see 3 to 8 below)	100 % of total current demand up to 10 amperes +50 % of any current demand in excess of 10 amperes	100 % f.l. of largest appliance +75 % f.l. of remaining appliances	100 % f.l. of largest appliance +80 % f.l. of second largest appliance +60 % f.l. of remaining appliances
3. Cooking appliances	10 amperes +30 % f.l. of connected cooking appliances in excess of 10 amperes +5 amperes if socket-outlet incorporated in control unit	100 % f.l. of largest appliance +80 % f.l. of second largest appliance +60 % f.l. of remaining appliances	100 % f.l. of largest appliance +80 % f.l. of second largest appliance +60 % f.l. of remaining appliances
4. Motors (other than lift motors which are subject to special consideration	not applicable	100 % f.l. of largest motor +80 % f.l. of second largest motor +60 % f.l. of remaining motors	100 % f.l. of largest motor +50 % f.l. of remaining motors
5. Water-heaters (instantaneous type)*	100 % f.l. of largest appliance +100 % f.l. of second largest appliance +25 % f.l. of remaining appliances	100 % f.l. of largest appliance +100 % f.l. of second largest appliance +25 % f.l. of remaining appliances	100 % f.l. of largest appliance +100 % f.l. of second largest appliance +25 % f.l. of remaining appliances

TABLE 1B continued Allowances for diversity

Purpose of final circuit fed from conductors or switchgear to which diversity applies	Type of premises		
	Individual household installations including individual dwellings of a block	Small shops, stores, offices and business premises	Small hotels, boarding houses, guest houses, etc
6. Water-heaters (thermostatically controlled)	no diversity allowable†		
7. Floor warming installations	no diversity allowable†		
8. Thermal storage space heating installations	no diversity allowable†		
9. Standard arrangement of final circuits in accordance with Appendix 8	100 % of current demand of largest circuit +40 % of current demand of every other circuit	100 % of current demand of largest circuit +50 % of current demand of every other circuit	
10. Socket-outlets other than those included in 9 above and stationary equipment other than those listed above	100 % of current demand of largest point of utilisation +40 % of current demand of every other point of utilisation	100 % of current demand of largest point of utilisation +70 % of current demand of every other point of utilisation	100 % of current demand of largest point of utilisation +75 % of current demand of every other point in main rooms (dining rooms, etc) +40 % of current demand of every other point of utilisation

* For the purpose of this Table an instantaneous water-heater is deemed to be a water-heater of any loading which heats water only while the tap is turned on and therefore uses electricity intermittently.

† It is important to ensure that the distribution boards and consumer units are of sufficient rating to take the total load connected to them without the application of any diversity.

APPENDIX 2

MAXIMUM PERMISSIBLE MEASURED EARTH FAULT LOOP IMPEDANCE

The tables in this Appendix provide maximum permissible 713-11 measured earth fault loop impedances (Zs) for compliance with 413-02-05 BS 7671 where the conventional final circuits of Table 7.1 are 413-02-10 used. The values are those that must not be exceeded in the 413-02-11 tests carried out under Para 10.3.6 at an ambient temperature 413-02-14 of 10 °C to 20 °C. Table 2E provides correction factors for other 543-01-03 ambient temperatures.

Where the cables to be used are to Table 5 or Table 6 of BS 6004 or Table 7 of BS 7211 or are other thermoplastic (pvc) or thermosetting (lsf) cables to these British Standards, and the cable loading is such that the maximum operating temperature is 70 °C, then Tables 2A, 2B and 2C give the maximum earth loop impedances for circuits with:

(a)   protective conductors of copper and having from 1 mm^2 to 16 mm^2 cross-sectional area,

(b)   where the overcurrent protective device is a fuse to BS 88 Part 2 or Part 6, BS 1361 or BS 3036.

For each type of fuse, two tables are given:

- where the circuit concerned feeds socket-outlets and the disconnection time for compliance with Regulation 413-02-09 is 0.4 s, and 413-02-09

- where the circuit concerned feeds fixed equipment only and the disconnection time for compliance with Regulation 413-02-13 is 5 s. 413-02-13

In each table the earth fault loop impedances given correspond to the appropriate disconnection time from a comparison of the time/current characteristic of the device concerned and the equation given in Regulation 543-01-03. 543-01-03

The tabulated values apply only when the nominal voltage to Earth (U$_o$) is 230 V.

Table 2D gives the maximum measured Z$_s$ for circuits protected by circuit-breakers to BS 3871-1 and BS EN 60898, and RCBOs to BS EN 61009.

Note: The impedances tabulated in this Appendix are lower than those in Table 41B1, Table 41B2 and Table 41D of BS 7671 as these are measured values at an assumed conductor temperature of 10 °C, whilst those in BS 7671 are design figures at the conductor normal operating temperature.

TABLE 2A Semi enclosed fuses
Maximum measured earth fault loop impedance (in ohms) when overcurrent protective device is a semi-enclosed fuse to BS 3036 (see Note)

(i)   0.4 second disconnection

413-02-05
Table 41B1
543-01-03

Protective conductor (mm^2)	Fuse rating (amperes)				
	5	15	20	30	45
1.0	8.00	2.14	1.48	NP	NP
1.5	8.00	2.14	1.48	0.91	NP
2.5 to 16.0	8.00	2.14	1.48	0.91	0.50

(ii)   5 seconds disconnection

413-02-05
Table 41D
543-01-03

Protective conductor (mm^2)	Fuse rating (amperes)				
	5	15	20	30	45
1.0	14.80	4.46	2.79	NP	NP
1.5	14.80	4.46	3.20	2.08	NP
2.5	14.80	4.46	3.20	2.21	1.20
4.0 to 16.0	14.80	4.46	3.20	2.21	1.33

Note: A value of k of 115 from Table 54C of BS 7671 is used. This is suitable for pvc insulated and sheathed cables to Table 5 or Table 6 of BS 6004 and for lsf insulated and sheathed cables to Table 7 of BS 7211. The k value is based on both the thermoplastic (pvc) and thermosetting (lsf) cables operating at a maximum temperature of 70 °C.

Table 54C

NP protective conductor, fuse combination NOT PERMITTED.

TABLE 2B BS 88 fuses
Maximum measured earth fault loop impedance (in ohms) when overcurrent protective device is a fuse to BS 88

(i)   0.4 second disconnection   **(see Note)**

413-02-05
Table 41B1
543-01-03

Protective conductor (mm²)	Fuse rating (amperes)							
	6	10	16	20	25	32	40	50
1.0	7.11	4.26	2.26	1.48	1.20	0.69	NP	NP
1.5	7.11	4.26	2.26	1.48	1.20	0.87	0.67	NP
2.5 to 16.0	7.11	4.26	2.26	1.48	1.20	0.87	0.69	0.51

(ii)   5 seconds disconnection

413-02-05
Table 41D
543-01-03

Protective conductor (mm²)	Fuse rating (amperes)							
	6	10	16	20	25	32	40	50
1.0	11.28	6.19	3.20	1.75	1.24	0.69	NP	NP
1.5	11.28	6.19	3.49	2.43	1.60	1.12	0.67	NP
2.5	11.28	6.19	3.49	2.43	1.92	1.52	1.13	0.56
4.0	11.28	6.19	3.49	2.43	1.92	1.52	1.13	0.81
6.0 to 16.0	11.28	6.19	3.49	2.43	1.92	1.52	1.13	0.87

Note: A value of k of 115 from Table 54C of BS 7671 is used. This is suitable for pvc insulated and sheathed cables to Table 5 or Table 6 of BS 6004 and for lsf insulated and sheathed cables to Table 7 of BS 7211. The k value is based on both the thermoplastic (pvc) and thermosetting (lsf) cables operating at a maximum temperature of 70 °C.

Table 54C

NP protective conductor, fuse combination NOT PERMITTED.

TABLE 2C
Maximum measured earth fault loop impedance (in ohms) when overcurrent protective device is a fuse to BS 1361

(i)   0.4 second disconnection

(see Note)

413-02-05
Table 41B1
543-01-03

Protective conductor (mm²)	Fuse rating (amperes)				
	5	15	20	30	45
1.0	8.72	2.74	1.42	0.80	NP
1.5	8.72	2.74	1.42	0.96	0.34
2.5 to 16.0	8.72	2.74	1.42	0.96	0.48

(ii)   5 seconds disconnection

Table 41D

Protective conductor (mm²)	Fuse rating (amperes)				
	5	15	20	30	45
1.0	13.68	4.18	1.75	0.80	NP
1.5	13.68	4.18	2.24	1.20	0.34
2.5	13.68	4.18	2.34	1.54	0.53
4.0	13.68	4.18	2.34	1.54	0.70
6.0 to 16.0	13.68	4.18	2.34	1.54	0.80

Note: A value of k of 115 from Table 54C of BS 7671 is used. This is suitable for pvc insulated and sheathed cables to Table 5 or Table 6 of BS 6004 and for lsf insulated and sheathed cables to Table 7 of BS 7211. The k value is based on both the thermoplastic (pvc) and thermosetting (lsf) cables operating at a maximum temperature of 70 °C.   Table 54C

NP protective conductor, fuse combination NOT PERMITTED.

TABLE 2D

Maximum measured earth fault loop impedance (in ohms) when overcurrent protective device is a circuit-breaker to BS 3871-1 or BS EN 60898 or a RCBO to BS EN 61009

Table 41B2
413-02-05

(i) both 0.4 and 5 seconds disconnection times

Circuit-breaker type	Circuit-breaker (amperes)												
	5	6	10	15	16	20	25	30	32	40	45	50	63
1	9.60	8.00	4.80	3.20	3.00	2.40	1.92	1.60	1.50	1.20	1.06	0.96	0.76
2	5.49	4.57	2.74	1.83	1.71	1.37	1.10	0.91	0.86	0.69	0.61	0.55	0.43
B	—	6.40	3.84	—	2.40	1.92	1.54	—	1.20	0.96	0.86	0.77	0.61
3&C	3.84	3.20	1.92	1.28	1.20	0.96	0.77	0.64	0.60	0.48	0.42	0.38	0.30
D	1.92	1.60	0.96	0.64	0.60	0.48	0.38	0.32	0.30	0.24	0.22	0.19	0.15

Note: A value of k of 115 from Table 54C of BS 7671 is used. This is suitable for thermoplastic (pvc) insulated and sheathed cables to Table 5 or Table 6 of BS 6004 and for thermosetting (lsf) insulated and sheathed cables to Table 7 of BS 7211. The k value is based on both the pvc and lsf cables operating at a maximum temperature of 70 °C.

TABLE 2E

Ambient temperature correction factors

Ambient temperature °C	Correction factors (from 10 °C) notes 1, 2
0	0.96
5	0.98
10	1.00
20	1.04
25	1.06
30	1.08

Notes:
1 - The correction factor is given by: {1 + 0.004 (Ambient temp - 10} where 0.004 is the simplified resistance coefficient per °C at 20 °C given by BS 6360 for both copper and aluminium conductors

2 - The factors are different to those of Table 9B because Table 2E corrects from 10 °C and Table 9B from 20 °C. The values in Tables 2A to 2D are for a 10 °C ambient.

The ambient correction factor of Table 2E is applied to the earth fault loop impedances of Tables 2A to 2D if the ambient temperature is not within the range 10 °C, to 20 °C. For example, if the ambient temperature is 25 °C the measured earth fault loop impedance of a circuit protected by a 32 A type 1 mcb should not exceed 1.50 x 1.06 = 1.59 Ω.

APPENDIX 3

NOTES ON THE SELECTION OF TYPES OF CABLE AND FLEXIBLE CORD FOR PARTICULAR USES AND EXTERNAL INFLUENCES

For compliance with the requirements of Chapter 52 for the Ch 52 selection and erection of wiring systems in relation to risks of mechanical damage and corrosion, this Appendix lists in two tables types of cable and flexible cord suitable for the uses intended. These tables are not intended to be exhaustive and other limitations may be imposed by the relevant Regulation of BS 7671, in particular those concerning maximum permissible operating temperatures.

Information is also included in this Appendix on protection against corrosion of exposed metalwork of wiring systems.

TABLE 3A
Applications of cables for fixed wiring

Type of cable	Uses	Comments
Thermoplastic (pvc) or thermosetting insulated non-sheathed	In conduits, cable ducting or trunking	(i) intermediate support may be required on long vertical runs (ii) 70 °C maximum conductor temperature for normal wiring grades — including thermosetting types (4) (iii)cables run in pvc conduit shall not operate with a conductor temperature greater than 70 °C (4)
Flat thermoplastic (pvc) or thermosetting, insulated and sheathed	(i) general indoor use in dry or damp locations. May be embedded in plaster (ii) on exterior surface walls, boundary walls and the like (iii)overhead wiring between buildings (6) (iv)underground in conduits or pipes (v) in building voids or ducts formed in situ	(i) additional protection may be necessary where exposed to mechanical stresses (ii) protection from direct sunlight may be necessary. Black sheath colour is better for cables in sunlight (iii)see Note (4) (iv)unsuitable for embedding directly in concrete (v) may need to be hard drawn (HD) copper conductors for overhead wiring (Note 6)

TABLE 3A continued
Applications of cables for fixed wiring

Type of cable	Uses	Comments
Split-concentric thermosetting (pvc) insulated and sheathed	General	(i) additional protection may be necessary where exposed to mechanical stresses (ii) protection from direct sunlight may be necessary. Black sheath colour is better for cables in sunlight
Mineral insulated	General	With overall pvc covering where exposed to the weather or risk of corrosion, or where installed underground, or in concrete ducts
Thermoplastic or thermosetting insulated, armoured, thermoplastic sheathed	General	(i) additional protection may be necessary where exposed to mechanical stresses (ii) protection from direct sunlight may be necessary. Black sheath colour is better for cables in sunlight
Paper insulated, lead sheathed and served	General, for main distribution cables	With armouring where exposed to severe mechanical stresses or where installed underground

Notes:

1 - The use of cable covers (preferably conforming to BS 2484) or equivalent mechanical protection is desirable for all underground cables which might otherwise subsequently be disturbed. Route marker tape should also be installed, buried just below ground level.

2 - Cables having thermoplastic (pvc) insulation or sheath should preferably not be used where the ambient temperature is consistently below 0 °C or has been within the preceding 24 hours. Where they are to be installed during a period of low temperature, precautions should be taken to avoid risk of mechanical damage during handling. A minimum ambient temperature of 5 °C is advised in BS 7540 : 1994 for some types of pvc insulated and sheathed cables

3 - Cables must be suitable for the maximum ambient temperature, and shall be protected from any excess heat produced by other equipment, including other cables.

4 - Thermosetting cable types (to BS 7211 or BS 5467) can operate with a conductor temperature of 90 °C. This must be limited to 70 °C when drawn into a conduit etc. with thermoplastic (pvc) insulated conductor (521-07-03) or connected to electrical equipment (512-02-01 and 523-01-01), or when such cables are installed in plastic conduit or trunking.

5 - For cables to BS 6004, BS 6007, BS 7211, BS 6346, BS 5467 and BS 6724, further guidance may be obtained from those standards. Additional advice is given in BS 7540 : 1994 "Guide to use of cables with a rated voltage not exceeding 450/750 V" for cables to BS 6004, BS 6007 and BS 7211.

6 - Cables for overhead wiring between buildings must be able to support their own weight and any imposed wind or ice/snow loading. A catenary support is usual but hard drawn copper types may be used.

Migration of plasticiser from thermoplastic (pvc) materials

Thermoplastic (pvc) sheathed cables, including thermosetting insulated with thermoplastic sheath e.g. lsf, must be separated from expanded polystyrene materials to prevent take up of the cable plasticiser by the polystyrene as this will reduce the flexibility of the cables.

Thermal insulation

Thermoplastic (pvc) sheathed cables in roof spaces must be clipped clear of any insulation made of expanded polystyrene granules.

Cable clips

Polystyrene cable clips are softened by contact with thermoplastic (pvc). Nylon and polypropylene are unaffected.

Grommets

Natural rubber grommets can be softened by contact with thermoplastic (pvc). Synthetic rubbers are more resistant. Thermoplastic (pvc) grommets are not affected, but could affect other plastics.

Wood preservatives

Thermoplastic (pvc) sheathed cables should be covered to prevent contact with preservative fluids during application. After the solvent has evaporated (good ventilation is necessary) the preservative has no effect.

Creosote

Creosote should not be applied to thermoplastic (pvc) sheathed cables because it causes decomposition, solution, swelling and loss of pliability.

TABLE 3B
Applications of flexible cables and cords to BS 6500 : 2000 and BS 7919 : 2001 generally

Type of flexible cord	Uses
Light thermoplastic (pvc) insulated and sheathed	Indoors in household or commercial premises in dry situations, for light duty
Ordinary thermoplastic (pvc) insulated and sheathed	(i) indoors in household or commercial premises, including damp situations, for medium duty (ii) for cooking and heating appliances where not in contact with hot parts (iii) for outdoor use other than in agricultural or industrial applications (iv) electrically powered hand tools
60 °C thermosetting (rubber) insulated braided twin and three-core	Indoors in household or commercial premises where subject only to low mechanical stresses
60 °C thermosetting (rubber) insulated and sheathed	(i) indoors in household or commercial premises where subject only to low mechanical stresses (ii) occasional use outdoors (iii) electrically powered hand tools
60 °C thermosetting (rubber) insulated oil-resisting and flame retardant sheath	(i) general, unless subject to severe mechanical stresses (ii) fixed installations protected in conduit or other enclosure
85 °C thermosetting (rubber) insulated HOFR sheathed	General, including hot situations, e.g. night storage heaters and immersion heaters
85 °C heat resisting thermoplastic (pvc) insulated and sheathed	General, including hot situations, e.g. for pendant luminaires
150 °C thermosetting (rubber) insulated and braided	(i) at high ambient temperatures (ii) in or on luminaires
185 °C glass fibre insulated single-core twisted twin and three-core	For internal wiring of luminaires only and then only where permitted by BS 4533
185 °C glass fibre insulated braided circular	(i) dry situations at high ambient temperatures and not subject to abrasions or undue flexing (ii) wiring of luminaires

Notes:

1 - Cables and cords having thermoplastic (pvc) insulation or sheath should preferably not be used where the ambient temperature is consistently below 0 °C. Where they are to be installed during a period of low temperature, precautions should be taken to avoid risk of mechanical damage during handling.

2 - Cables and cords shall be suitable for the maximum ambient temperature, and shall be protected from any excess heat produced by other equipment, including other cables.

3 - For flexible cords and cables to BS 6007, BS 6141 and BS 6500 further guidance may be obtained from those standards, or from BS 7540 : 1994 "Guide to use of cables with a rated voltage not exceeding 450/750 V".

4 - When used as connections to equipment flexible cables and cords should be of the minimum practical length to minimise danger and in any case of such a length that allows the protective device to operate correctly.

5 - When attached to equipment flexible cables and cords should be protected against tension, crushing, abrasion, torsion and kinking particularly at the inlet point to the electrical equipment. At such inlet points it may be necessary to use a device which ensures that the cable is not bent to an internal radius below that given in the appropriate part of Table 4 of BS 6700. Strain relief, clamping devices or cord guards should not damage the cord.

6 - Flexible cables and cords should not be used under carpets or other floor coverings, or where furniture or other equipment may rest on them. Flexible cables and cords should not be placed where there is a risk of damage from traffic passing over them.

7 - Flexible cables and cords should not be used in contact with or close to heated surfaces, especially if the surface approaches the upper thermal limit of the cable or cord.

Protection against corrosion of exposed metalwork or wiring systems   522

In damp situations, where metal cable sheaths and armour of 522-03 cables, metal conduit and conduit fittings, metal ducting and 522-05 trunking systems, and associated metal fixings, are liable to chemical deterioration or electrolytic attack by materials of a structure with which they may come in contact, it is necessary to take suitable precautions against corrosion.

Materials likely to cause such attack include:

- materials containing magnesium chloride which are used in the construction of floors and dadoes,

- plaster undercoats contaminated with corrosive salts,

- lime, cement and plaster, for example on unpainted walls,

- oak and other acidic woods,

- dissimilar metals likely to set up electrolytic action.

Application of suitable coatings before erection, or prevention of contact by separation with plastics, are recognized as effective precautions against corrosion.

Special care is required in the choice of materials for clips and other fittings for bare aluminium sheathed cables and for aluminium conduit, to avoid risk of local corrosion in damp situations. Examples of suitable materials for this purpose are the following: 522-05-02 522-05-03

- porcelain,

- plastics,

- aluminium,

- corrosion-resistant aluminium alloys,

- zinc alloys complying with BS 1004,

- iron or steel protected against corrosion by galvanizing, sherardizing etc.

Contact between bare aluminium sheaths or aluminium conduits and any parts made of brass or other metal having a high copper content should be especially avoided in damp situations, unless the parts are suitably plated. If such contact is unavoidable, the joint should be completely protected against ingress of moisture. Wiped joints in aluminium sheathed cables should always be protected against moisture by a suitable paint, by an impervious tape, or by embedding in bitumen. 522-05-02

NOTES ON METHODS OF SUPPORT FOR CABLES, CONDUCTORS AND WIRING SYSTEMS

This Appendix describes examples of methods of support for cables, conductors and wiring systems which should satisfy the relevant requirements of Chapter 52 of BS 7671. The use of other methods is not precluded where specified by a suitably qualified electrical engineer.

Cables generally

Items 1 to 8 below are generally applicable to supports on structures which are subject only to vibration of low severity and a low risk of mechanical impact.

1. For non-sheathed cables, installation in conduit without further fixing of the cables, precautions being taken against undue compression or other mechanical stressing of the insulation at the top of any vertical runs exceeding 5 m in length.

2. For cables of any type, installation in ducting or trunking without further fixing of the cables, vertical runs not exceeding 5 m in length without intermediate support.

3. For sheathed and/or armoured cables installed in accessible positions, support by clips at spacings not exceeding the appropriate value stated in Table 4A.

4. For cables of any type, resting without fixing in horizontal runs of ducts, conduits, cable ducting or trunking.

5. For sheathed and/or armoured cables in horizontal runs which are inaccessible and unlikely to be disturbed, resting without fixing on part of a building, the surface of that part being reasonably smooth.

6. For sheathed-and-armoured cables in vertical runs which are inaccessible and unlikely to be disturbed, supported at the top of the run by a clip and a rounded support of a radius not less than the appropriate value stated in Table 4E.

7.   For sheathed cables without armour in vertical runs which are inaccessible and unlikely to be disturbed, supported by the method described in Item 6 above; the length of run without intermediate support not exceeding 2 m for a lead sheathed cable or 5 m for a thermosetting or thermoplastic sheathed cable.

8.   For thermosetting or thermoplastic (pvc) sheathed cables, installation in conduit without further fixing of the cables, any vertical runs being in conduit of suitable size and not exceeding 5 m in length.

Cables in particular conditions

9.   In caravans, for sheathed cables in inaccessible spaces such as ceiling, wall and floor spaces, support at intervals not exceeding 0.25 m for horizontal runs and 0.4 m for vertical runs.

10.   In caravans, for horizontal runs of sheathed cables passing through floor or ceiling joists in inaccessible floor or ceiling spaces, securely bedded in thermal insulating material, no further fixing is required.

11.   For flexible cords used as pendants, attachment to a ceiling rose or similar accessory by the cord grip or other method of strain relief provided in the accessory.

12.   For temporary installations and installations on construction sites, supports so arranged that there is no appreciable mechanical strain on any cable termination or joint.

Overhead wiring

13.   For cables sheathed with thermosetting or thermoplastic material, supported by a separate catenary wire, either continuously bound up with the cable or attached thereto at intervals; the intervals not exceeding those stated in Column 2 of Table 4A.

14.   Support by a catenary wire incorporated in the cable during manufacture, the spacings between supports not exceeding those stated by the manufacturer and the minimum height above ground being in accordance with Table 4B.

15.   For spans without intermediate support (e.g. between buildings) of thermoplastic (pvc)-insulated thermoplastic (pvc)-sheathed cable, or thermosetting-insulated cable having an oil-resisting and flame-retardant or HOFR sheath, terminal supports so arranged that no undue strain is placed upon the conductors or insulation of the cable, adequate precautions being taken against any risk of chafing of the cable sheath, and the minimum height above ground and the length of such spans being in accordance with the appropriate values indicated in Table 4B.

16.   Bare or thermoplastic (pvc)-covered conductors of an overhead line for distribution between a building and a remote point of utilisation (e.g. another building) supported on insulators, the lengths of span and heights above ground having the appropriate values indicated in Table 4B or otherwise installed in accordance with the Electricity Supply Regulations 1988 (as amended).

17.   For spans without intermediate support (e.g. between buildings) and which are in situations inaccessible to vehicular traffic, cables installed in heavy gauge steel conduit, the length of span and height above ground being in accordance with Table 4B.

Conduit and cable trunking

18.   Rigid conduit supported in accordance with Table 4C.

19.   Cable trunking supported in accordance with Table 4D.

20.   Conduit embedded in the material of the building.

21.   Pliable conduit embedded in the material of the building or in the ground, or supported in accordance with Table 4C.

TABLE 4A
Spacings of supports for cables in accessible positions

Maximum spacings of clips

Overall diameter of cable*	Non-armoured thermosetting, thermoplastic or lead sheathed cables				Armoured cables		Mineral insulated copper sheathed or aluminium sheathed cables	
	Generally		In caravans					
	Horizontal† 2	Vertical† 3	Horizontal† 4	Vertical† 5	Horizontal† 6	Vertical† 7	Horizontal† 8	Vertical† 9
mm	mm	mm	mm	mm	mm	mm	mm	mm
Not exceeding 9	250	400	250 (for all sizes)	400 (for all sizes)	—	—	600	800
Exceeding 9 and not exceeding 15	300	400			350	450	900	1200
Exceeding 15 and not exceeding 20	350	450			400	550	1500	2000
Exceeding 20 and not exceeding 40	400	550			450	600	—	—

Note: For the spacing of supports for cables having an overall diameter exceeding 40 mm, and for single-core cables having conductors of cross-sectional area 300 mm^2 and larger, the manufacturer's recommendations should be observed.

* For flat cables taken as the dimension of the major axis.

† The spacings stated for horizontal runs may be applied also to runs at an angle of more than 30 from the vertical. For runs at an angle of 30° or less from the vertical, the vertical spacings are applicable.

TABLE 4B

Maximum lengths of span and minimum heights above ground for overhead wiring between buildings etc

Type of system 1	Maximum length of span 2	At road crossings 3	In positions accessible to vehicular traffic, other than crossings 4	In positions inaccessible to vehicular traffic* 5
		Minimum height of span above ground		
	m	m	m	m
Cables sheathed with thermoplastic (pvc) or having an oil-resisting and flame-retardant or HOFR sheath, without intermediate support. (Item 15)	3	(5.8 for all types)	(5.2 for all types)	3.5
Cables sheathed with thermoplastic (pvc) or having an oil-resisting and flame-retardant or HOFR sheath, in heavy gauge steel conduit of diameter not less than 20 mm and not jointed in its span. (Item 17)	3			3
Bare or thermoplastic (pvc) covered overhead lines on insulators without intermediate support. (Item 16)	30			3.5
Cables sheathed with thermoplastic (pvc) or having an oil-resisting and flame-retardant or HOFR sheath, supported by a catenary wire. (Item 13)	No limit			3.5

TABLE 4B continued
Maximum lengths of span and minimum heights above ground for overhead wiring between buildings etc

Type of system 1	Maximum length of span 2	Minimum height of span above ground		
		At road crossings 3	In positions accessible to vehicular traffic, other than crossings 4	In positions inaccessible to vehicular traffic * 5
	m	m	m	m
Aerial cables incorporating a catenary wire. (Item 14)	Subject to Item 14	(5.8 for all types)	(5.2 for all types)	3.5
Bare or thermoplastic (pvc) covered overhead lines installed in accordance with the Overhead Line Regulations (Item 16).	No limit			5.2

*   This column is not applicable in agricultural premises.

Note: In some special cases, such as in yacht marinas or where large cranes are present, it will be necessary to increase the minimum height of span above ground given in Table 4B. It is preferable to use underground cables in such locations.

TABLE 4C
Spacings of supports for conduits

Nominal size of conduit	Maximum distance between supports					
	Rigid metal		Rigid insulating		Pliable	
	Horizontal	Vertical	Horizontal	Vertical	Horizontal	Vertical
1	2	3	4	5	6	7
mm	m	m	m	m	m	m
Not exceeding 16	0.75	1.0	0.75	1.0	0.3	0.5
Exceeding 16 and not exceeding 25	1.75	2.0	1.5	1.75	0.4	0.6
Exceeding 25 and not exceeding 40	2.0	2.25	1.75	2.0	0.6	0.8
Exceeding 40	2.25	2.5	2.0	2.0	0.8	1.0

Notes:

1. The spacings tabulated allow for maximum fill of cables permitted by the Regulations and the thermal limits specified in the relevant British Standards. They assume that the conduit or trunking is not exposed to other mechanical stress.

2. The above figures do not apply to lighting suspension trunking, where the manufacturer's instructions must be followed, or where special strengthening couplers are used. A flexible conduit is not normally required to be supported in its run. Supports should be positioned within 300 mm of bends or fittings.

3. A flexible conduit should be of such length that it does not need to be supported in its run.
The inner radius of a conduit bend should be not less than 2.5 times the outside diameter of the conduit.

TABLE 4D
Spacings of supports for cable trunking

Cross-sectional area of trunking	Maximum distance between supports			
	metal		insulating	
	Horizontal	Vertical	Horizontal	Vertical
1	2	3	4	5
mm²	m	m	m	m
Exceeding 300 and not exceeding 700	0.75	1.0	0.5	0.5
Exceeding 700 and not exceeding 1500	1.25	1.5	0.5	0.5
Exceeding 1500 and not exceeding 2500	1.75	2.0	1.25	1.25
Exceeding 2500 and not exceeding 5000	3.0	3.0	1.5	2.0
Exceeding 5000	3.0	3.0	1.75	2.0

Notes:

1. The spacings tabulated allow for maximum fill of cables permitted by the Regulations and the thermal limits specified in the relevant British Standards. They assume that the conduit or trunking is not exposed to other mechanical stress.

2. The above figures do not apply to lighting suspension trunking, where the manufacturer's instructions must be followed, or where special strengthening couplers are used. A flexible conduit is not normally required to be supported in its run. Supports should be positioned within 300 mm of bends or fittings.

TABLE 4E
Minimum internal radii of bends in cables for fixed wiring

Insulation	Finish	Overall diameter*	Factor to be applied to overall diameter of cable to determine minimum internal radius of bend
Thermosetting or thermoplastic (pvc) (circular, or circular stranded copper or aluminium conductors)	Non-armoured	Not exceeding 10 mm	3(2)†
		Exceeding 10 mm but not exceeding 25 mm	4(3)†
		exceeding 25 mm	6
	Armoured	Any	6
Thermosetting or thermoplastic (pvc) (solid aluminium or shaped copper conductors)	Armoured or non-armoured	Any	8
Mineral	Copper sheath with or without covering	Any	6‡

* For flat cables the diameter refers to the major axis.

† The figure in brackets relates to single-core circular conductors of stranded construction installed in conduit, ducting or trunking.

‡ Mineral insulated cables may be bent to a radius not less than 3 times the cable diameter over the copper sheath, provided that the bend is not re-worked, i.e. straightened and re-bent.

APPENDIX 5

CABLE CAPACITIES OF CONDUIT AND TRUNKING

A number of variable factors affect any attempt to arrive at a standard method of assessing the capacity of conduit or trunking. 522-08-01 522-08-02 522-08-03

Some of these are:

- reasonable care (of drawing-in)
- acceptable use of the space available
- tolerance in cable sizes
- tolerance in conduit and trunking.

The following tables can only give guidance of the maximum number of cables which should be drawn in. The sizes should ensure an easy pull with low risk of damage to the cables.

Only the ease of drawing-in is taken into account. The electrical effects of grouping are not. As the number of circuits increases the installed current-carrying capacity of the cable decreases. Cable sizes have to be increased with consequent increase in cost of cable and conduit.

It may therefore be more attractive economically to divide the circuits concerned between two or more enclosures.

If thermosetting cables are installed in the same conduit or trunking as thermoplastic (pvc) insulated cables, the conductor operating temperature of any of the cables must not exceed that for thermoplastic (pvc) i.e. thermosetting cables must be rated as thermoplastic (pvc).

The following three cases are dealt with:

Single-core thermoplastic (pvc) insulated cables

(i)   in straight runs of conduit not exceeding 3 m in length. Tables 5A & 5B

(ii)   in straight runs of conduit exceeding 3 m in length, or in runs of any length incorporating bends or sets. Tables 5C & 5D

(iii)   in trunking. Tables 5E & 5F.

For cables and/or conduits, not covered by this Appendix advice on the number of cables which can be drawn in should be obtained from the manufacturers.

Single-core thermoplastic (pvc) insulated cables in straight runs of conduit not exceeding 3 m in length.

For each cable it is intended to use, obtain the factor from Table 5A.

Add the cable factors together and compare the total with the conduit factors given in Table 5B.

The minimum conduit size is that having a factor equal to or greater than the sum of the cable factors.

<table>
<tr><td colspan="3">TABLE 5A
Cable factors for use in conduit in short straight runs</td><td colspan="2">TABLE 5B
Conduit factors for use in short straight runs</td></tr>
<tr><td>Type of conductor</td><td>Conductor cross-sectional area mm^2</td><td>Cable factor</td><td>Conduit diameter mm</td><td>Conduit factor</td></tr>
<tr><td></td><td></td><td></td><td>16</td><td>290</td></tr>
<tr><td></td><td>1</td><td>22</td><td></td><td></td></tr>
<tr><td>Solid</td><td>1.5</td><td>27</td><td>20</td><td>460</td></tr>
<tr><td></td><td>2.5</td><td>39</td><td></td><td></td></tr>
<tr><td></td><td></td><td></td><td>25</td><td>800</td></tr>
<tr><td></td><td>1.5</td><td>31</td><td>32</td><td>1400</td></tr>
<tr><td></td><td>2.5</td><td>43</td><td></td><td></td></tr>
<tr><td>Stranded</td><td>4</td><td>58</td><td>38</td><td>1900</td></tr>
<tr><td></td><td>6</td><td>88</td><td></td><td></td></tr>
<tr><td></td><td>10</td><td>146</td><td>50</td><td>3500</td></tr>
<tr><td></td><td>16</td><td>202</td><td></td><td></td></tr>
<tr><td></td><td>25</td><td>385</td><td>63</td><td>5600</td></tr>
</table>

Single-core thermoplastic (pvc)-insulated cables in straight runs of conduit exceeding 3 m in length or in runs of any length incorporating bends or sets.

For each cable it is intended to use, obtain the appropriate factor from Table 5C.

Add the cable factors together and compare the total with the conduit factors given in Table 5D, taking into account the length of run it is intended to use and the number of bends and sets in that run.

The minimum conduit size is that size having a factor equal to or greater than the sum of the cable factors. For the larger sizes of conduit multiplication factors are given relating them to 32 mm diameter conduit.

TABLE 5C

Cable factors for use in conduit in long straight runs over 3 m, or runs of any length incorporating bends

Type of conductor	Conductor cross-sectional area mm^2	Cable factor
Solid or Stranded	1	16
	1.5	22
	2.5	30
	4	43
	6	58
	10	105
	16	145
	25	217

The inner radius of a conduit bend should be not less than 2.5 times the outside diameter of the conduit.

TABLE 5D
Cable factors for runs incorporating bends and long straight runs

Conduit diameter, mm

length of run m	Straight 16	20	25	32	One bend 16	20	25	32	Two bends 16	20	25	32	Three bends 16	20	25	32	Four bends 16	20	25	32
1					188	303	543	947	177	286	514	900	158	256	463	818	130	213	388	692
1.5	Covered by				182	294	528	923	167	270	487	857	143	233	422	750	111	182	333	600
2	Tables				177	286	514	900	158	256	463	818	130	213	388	692	97	159	292	529
2.5	A and B				171	278	500	878	150	244	442	783	120	196	358	643	86	141	260	474
3					167	270	487	857	143	233	422	750	111	182	333	600				
3.5	179	290	521	911	162	263	475	837	136	222	404	720	103	169	311	563				
4	177	286	514	900	158	256	463	818	130	213	388	692	97	159	292	529				
4.5	174	282	507	889	154	250	452	800	125	204	373	667	91	149	275	500				
5	171	278	500	878	150	244	442	783	120	196	358	643	86	141	260	474				
6	167	270	487	857	143	233	422	750	111	182	333	600								
7	162	263	475	837	136	222	404	720	103	169	311	563								
8	158	256	463	818	130	213	388	692	97	159	292	529								
9	154	250	452	800	125	204	373	667	91	149	275	500								
10	150	244	442	783	120	196	358	643	86	141	260	474								

Additional Factors: For 38 mm diameter use . . . . 1.4 x (32 mm factor)
For 50 mm diameter use . . . . 2.6 x (32 mm factor)
For 63 mm diameter use . . . . 4.2 x (32 mm factor)

Single-core thermoplastic (pvc)-insulated cables in trunking

For each cable it is intended to use, obtain the appropriate factor from Table 5E.

Add all the cable factors so obtained and compare with the factors for trunking given in Table 5F.

The minimum size of trunking is that size having a factor equal to or greater than the sum of the cable factors.

TABLE 5E
Cable factors for trunking

Type of conductor	Conductor cross-sectional area mm^2	PVC, BS 6004 Cable factor	Thermosetting BS 7211 Cable factor
Solid	1.5	8.0	8.6
	2.5	11.9	11.9
Stranded	1.5	8.6	9.6
	2.5	12.6	13.9
	4	16.6	18.1
	6	21.2	22.9
	10	35.3	36.3
	16	47.8	50.3
	25	73.9	75.4

Note:

(i)   These factors are for metal trunking and may be optimistic for plastic trunking where the cross-sectional area available may be significantly reduced from the nominal by the thickness of the wall material.

(ii)   The provision of spare space is advisable; however, any circuits added at a later date must take into account grouping. Appendix 4, BS 7671.

TABLE 5F
Factors for trunking

Dimensions of trunking mm x mm	Factor	Dimensions of trunking mm x mm	Factor
50 x 38	767	200 x 100	8572
50 x 50	1037	200 x 150	13001
75 x 25	738	200 x 200	17429
75 x 38	1146	225 x 38	3474
75 x 50	1555	225 x 50	4671
75 x 75	2371	225 x 75	7167
100 x 25	993	225 x 100	9662
100 x 38	1542	225 x 150	14652
100 x 50	2091	225 x 200	19643
100 x 75	3189	225 x 225	22138
100 x 100	4252	300 x 38	4648
150 x 38	2999	300 x 50	6251
150 x 50	3091	300 x 75	9590
150 x 75	4743	300 x 100	12929
150 x 100	6394	300 x 150	19607
150 x 150	9697	300 x 200	26285
200 x 38	3082	300 x 225	29624
200 x 50	4145	300 x 300	39428
200 x 75	6359		

Space factor - 45 % with trunking thickness taken into account

For other sizes and types of cable or trunking

For sizes and types of cable trunking other than those given in Tables 5E and 5F, the number of cables installed should be such that the resulting space factor does not exceed 45 % of the net internal cross-sectional area.

Space factor is defined as the ratio (expressed as a percentage) of the sum of the overall cross-sectional areas of cables (insulation and any sheath) to the internal cross-sectional area of the trunking or other cable enclosure in which they are installed. The effective overall cross-sectional area of a non-circular cable is taken as that of a circle of diameter equal to the major axis of the cable.

Care should be taken to use trunking bends etc which do not impose bending radii on cables less than those required by Table 4E.

CURRENT-CARRYING CAPACITIES AND VOLTAGE DROP FOR COPPER CONDUCTORS

Current-carrying Capacity

For full information on the selection of cables including calculation of voltage drop see Appendix 4 of BS 7671. 523-01-01 App 4

In this simplified approach it is presumed that the overcurrent device will be providing both fault and overload current protection.

Procedure

(1)   the design current I_b of the circuit must be established

(2)   the overcurrent device rating I_n is then selected so that I_n is greater than or equal to I_b 433-02-01 433-02-02

$$I_n \geq I_b$$

The tabulated current-carrying capacity of the selected cable I_t is then given by

$$I_t \geq \frac{I_n}{C_a \, C_i \, C_g \, C_r}$$

for simultaneously occurring factors.

Where :
C_a is the correction factor for ambient temperature, see Tables 6A1 and 6A2 App 4,4
C_i is the correction factor for thermal insulation, see Table 6B
C_g is the correction factor for grouping, see Table 6C
C_r is the correction factor 0.725 for semi-enclosed fuses to BS 3036 433-02-03

Voltage Drop

To calculate the voltage drop in volts the tabulated value of voltage drop (mV/A/m) has to be multiplied by the design App 4,7 current of the circuit (I_b), the length of run in metres (L), and divided by 1000 (to convert to volts)

$$voltage\ drop = \frac{(mV/A/m) \times I_b \times L}{1000}$$

The requirements of BS 7671 are deemed to be satisfied for a 230 V supply, if the voltage drop between the origin of the installation and a socket-outlet or fixed current-using equipment does not exceed 9.2 V at full load. 525-01-02

TABLE 6A1 Ambient Temperature Factors

Correction factors for ambient temperature where protection is against short-circuit and overload Table 4C1

Type of insulation	Operating temperature	Ambient temperature °C								
		25	30	35	40	45	50	55	60	65
Thermoplastic (general purpose pvc)	70 °C	1.03	1.0	0.94	0.87	0.79	0.71	0.61	0.50	0.35

Note: Where the device is a semi-enclosed fuse to BS 3036 the table only applies where the device is intended to provide short-circuit protection only.

TABLE 6A2 Ambient Temperature Factors

Correction factors for ambient temperature where the overload protective device is a semi-enclosed fuse to BS 3036 Table 4C2

Type of insulation	Operating temperature	Ambient temperature °C								
		25	30	35	40	45	50	55	60	65
Thermoplastic (general purpose pvc)	70 °C	1.03	1.0	0.97	0.94	0.91	0.87	0.84	0.69	0.48

Where a cable is to be run in a space to which thermal insulation is likely to be applied, the cable shall wherever practicable be fixed in a position such that it will not be covered by the thermal insulation. Where fixing in such a position is impracticable the cross-sectional area of the cable shall be appropriately increased.

For a cable installed in a thermally insulated wall or above a thermally insulated ceiling, the cable being in contact with a thermally conductive surface on one side, current-carrying capacities are tabulated in Tables 6D and 6E, Method 4 being the appropriate Reference Method; and Table 6F Installation Methods 6 and 15.

For a single cable likely to be totally surrounded by thermally insulating material over a length of more than 0.5 m, the current-carrying capacity shall be taken, in the absence of more precise information, as 0.5 times the current-carrying capacity for that cable clipped direct to a surface and open (Reference Method 1).

Where a cable is totally surrounded by thermal insulation for less than 0.5 m the current-carrying capacity of the cable shall be reduced appropriately depending on the size of cable length in insulation and thermal properties of the insulation. The derating factors in the table are appropriate to conductor sizes up to 10 mm^2 in thermal insulation having a thermal conductivity (λ) greater than 0.0625 Wm^{-1}K^{-1}.

TABLE 6B Thermal Insulation   Table 52A

Cables surrounded by thermal insulation

Length in insulation (mm)	Derating factor
50	0.89
100	0.81
200	0.68
400	0.55
500 and over	0.50

TABLE 6C Grouping Factors

Correction factors for groups of more than one circuit of single-core cables, or more than one multicore cable**

Reference method of installation		Correction factor (C_g) Number of circuits or multicore cables													
		2	3	4	5	6	7	8	9	10	12	14	16	18	20
Enclosed (Method 3 or4) or bunched and clipped direct to a non-metallic surface (Method 1)		0.80	0.70	0.65	0.60	0.57	0.54	0.52	0.50	0.48	0.45	0.43	0.41	0.39	0.38
Single layer clipped to a non-metallic surface (Method 1)	Touching	0.85	0.79	0.75	0.73	0.72	0.72	0.71	0.70	–	–	–	–	–	–
	Spaced*	0.94	0.90	0.90	0.90	0.90	0.90	0.90	0.90	0.90	0.90	0.90	0.90	0.90	0.90
Single layer *multicore* on a perforated metal cable tray, vertical or horizontal (Method 11)	Touching	0.86	0.81	0.77	0.75	0.74	0.73	0.73	0.72	0.71	0.70	–	–	–	–
	Spaced*	0.91	0.89	0.88	0.87	0.87	–	–	–	–	–	–	–	–	–
Single layer *single-core* on a perforated metal cable tray, touching (Method 11)	Horizontal	0.90	0.85	–	–	–	–	–	–	–	–	–	–	–	–
	Vertical	0.85	–	–	–	–	–	–	–	–	–	–	–	–	–
Single layer multicore touching on ladder supports (Method 13)		0.86	0.82	0.80	0.79	0.78	0.78	0.78	0.77	–	–	–	–	–	–

NOTES TO TABLE 6C.

* Spaced by a clearance between adjacent surfaces of at least one cable diameter D_e. Where the horizontal clearances between adjacent cables exceeds $2D_e$ no correction factor need be applied.

** When cables having differing conductor operating temperatures are grouped together, the current rating shall be based upon the lowest operating temperature of any cable in the group.

1. The factors in the table are applicable to groups of cables all of one size. The value of current derived from application of the appropriate factors is the maximum current to be carried by any of the cables in the group.

2. If, owing to known operating conditions, a cable is expected to carry not more than 30 % of its grouped rating, it may be ignored for the purpose of obtaining the rating factor for the rest of the group.

   For example, a group of N loaded cables would normally require a group reduction factor of C_g applied to the tabulated I_t. However, if M cables in the group carry loads which are not greater than 0.3 $C_g I_t$ amperes the other cables can be sized by using the group rating factor corresponding to (N-M) cables.

3. For mineral insulated cables see Table 4B2 of BS 7671.

— Correction factor not tabulated.

TABLE 6D1

Table 4D1A

Single-core cables having thermoplastic (pvc) or thermosetting insulation (note 1), non-armoured, with or without sheath

(COPPER CONDUCTORS)

Ambient temperature: 30 °C. Conductor operating temperature: 70 °C

CURRENT-CARRYING CAPACITY (Amperes): BS 6004

Conductor cross-sectional area	Reference Method 4 (enclosed in conduit in thermally insulating wall etc.)		Reference Method 3 (enclosed in conduit on a wall or in trunking etc.)		Reference Method 1 (clipped direct)		Reference Method 11 (on a perforated cable tray horizontal or vertical)		Reference Method 12 (free air)		
	2 cables, single-phase a.c. or d.c.	3 or 4 cables, three-phase a.c.	2 cables, single-phase a.c. or d.c.	3 or 4 cables, three-phase a.c.	2 cables, single-phase a.c. or d.c. flat and touching	3 or 4 cables, three-phase a.c. flat and touching or trefoil	2 cables, single-phase a.c. or d.c. flat and touching	3 or 4 cables, three-phase a.c. flat and touching or trefoil	Horizontal flat spaced 2 cables, single-phase a.c. or d.c. or 3 cables three-phase a.c.	Vertical flat spaced 2 cables, single-phase a.c. or d.c. or 3 cables three-phase a.c.	Trefoil 3 cables trefoil, three-phase a.c.
1	2	3	4	5	6	7	8	9	10	11	12
mm²	A	A	A	A	A	A	A	A	A	A	A
1	11	10.5	13.5	12	15.5	14	–	–	–	–	–
1.5	14.5	13.5	17.5	15.5	20	18	–	–	–	–	–
2.5	20	18	24	21	27	25	–	–	–	–	–
4	26	24	32	28	37	33	–	–	–	–	–
6	34	31	41	36	47	43	–	–	–	–	–
10	46	42	57	50	65	59	–	–	–	–	–
16	61	56	76	68	87	79	–	–	–	–	–
25	80	73	101	89	114	104	126	112	146	130	110
35	99	89	125	110	141	129	156	141	181	162	137
50	119	108	151	134	182	167	191	172	219	197	167
70	151	136	192	171	234	214	246	223	281	254	216
95	182	164	232	207	284	261	300	273	341	311	264

NOTES TO TABLE 6D1:

1. The ratings for cables with thermosetting insulation are applicable for cables connected to equipment or accessories designed to operate with cables which run at a temperature not exceeding 70 °C. Where conductor operating temperatures up to 90 °C are acceptable the current rating is increased - see Table 4E1A of BS 7671.

   Table 4E1A

2. Where the conductor is to be protected by a semi-enclosed fuse to BS 3036, see the introduction to this Appendix.

3. The current-carrying capacities in columns 2 to 5 are also applicable to flexible cables to BS 6004 Table 1(c) and to 90 °C heat resisting pvc cables to BS 6231 Tables 8 and 9 where the cables are used in fixed installations.

TABLE 6D2

voltage drop (per ampere per metre): Conductor operating temperature: 70 °C

Conductor cross-sectional area	2 cables d.c.	2 cables, single-phase a.c.			3 or 4 cables, three-phase a.c.			
		Reference Methods 3 & 4 (enclosed in conduit etc. in or on a wall)	Reference Methods 1 &11 (clipped direct or on trays, touching)	Reference Method 12 (spaced*)	Reference Methods 3 &4 (enclosed in conduit etc. in or on a wall)	Reference Methods 1, 11 & 12 (in trefoil)	Reference Methods 1 &11 (flat and touching)	Reference Method 12 (flat spaced*)
1	2	3	4	5	6	7	8	9
mm²	mV/A/m	mV/A/m	mV/A/m	mV/A/m	mV/A/m	mV/A/m	mV/A/m	mV/A/m
1	44	44	44	44	38	38	38	38
1.5	29	29	29	29	25	25	25	25
2.5	18	18	18	18	15	15	15	15
4	11	11	11	11	9.5	9.5	9.5	9.5
6	7.3	7.3	7.3	7.3	6.4	6.4	6.4	6.4
10	4.4	4.4	4.4	4.4	3.8	3.8	3.8	3.8
16	2.8	2.8	2.8	2.8	2.4	2.4	2.4	2.4
		r†	r†	r†	r†	r†	r†	r†
25	1.75	1.80	1.75	1.75	1.50	1.50	1.50	1.50
35	1.25	1.30	1.25	1.25	1.10	1.10	1.10	1.10
50	0.93	0.95	0.93	0.93	0.81	0.80	0.80	0.80
70	0.63	0.65	0.63	0.63	0.56	0.55	0.55	0.55
95	0.46	0.49	0.47	0.47	0.42	0.41	0.41	0.40

* Note: Spacings larger than those specified in Reference Method 12 (see notes to Table 6C) will result in larger voltage drop.

† Note: The reactive element of voltage drop usually provided for 25 mm² and above conductor size, is omitted for simplicity. For a fuller treatment see Appendix 4 of BS 7671.

TABLE 6E1

Table 4D2A

Multicore cables having thermoplastic (pvc) or thermosetting insulation (note 1), non-armoured, (COPPER CONDUCTORS)

Ambient temperature: 30 °C. Conductor operating temperature: 70 °C

CURRENT-CARRYING CAPACITY (Amperes): BS 6004, BS 7629

Conductor cross-sectional area	Reference Method 4 (enclosed in an insulated wall, etc.)		Reference Method 3 (enclosed in conduit on a wall or ceiling, or in trunking)		Reference Method 1 (clipped direct)		Reference Method 11 (on a perforated cable tray) or Reference Method 13 (free air)	
	1 two-core cable*, single-phase a.c. or d.c.	1 three-core cable* or 1 four-core cable, three-phase a.c.	1 two-core cable*, single-phase a.c. or d.c.	1 three-core cable* or 1 four-core cable, three-phase a.c.	1 two-core cable*, single-phase a.c. or d.c.	1 three-core cable* or 1 four-core cable, three-phase a.c.	1 two-core cable*, single-phase a.c. or d.c.	1 three-core cable* or 1 four-core cable, three-phase a.c.
1	2	3	4	5	6	7	8	9
mm²	A	A	A	A	A	A	A	A
1	11	10	13	11.5	15	13.5	17	14.5
1.5	14	13	16.5	15	19.5	17.5	22	18.5
2.5	18.5	17.5	23	20	27	24	30	25
4	25	23	30	27	36	32	40	34
6	32	29	38	34	46	41	51	43
10	43	39	52	46	63	57	70	60
16	57	52	69	62	85	76	94	80
25	75	68	90	80	112	96	119	101
35	92	83	111	99	138	119	148	126
50	110	99	133	118	168	144	180	153
70	139	125	168	149	213	184	232	196
95	167	150	201	179	258	223	282	238

See Notes overleaf

NOTES TO TABLE 6E1:

1. The ratings for cables with thermosetting insulation are applicable for cables connected to equipment or accessories designed to operate with cables which run at a temperature not exceeding 70 °C. Where conductor operating temperatures up to 90 °C are acceptable the current rating is increased - see Table 4E2A of BS 7671.

Table 4E2A

2. Where the conductor is to be protected by a semi-enclosed fuse to BS 3036, see the introduction to this Appendix.

*3. With or without protective conductor. Circular conductors are assumed for sizes up to and including 16 mm^2. Values for larger sizes relate to shaped conductors and may safely be applied to circular conductors.

TABLE 6E2

Voltage drop:
(per ampere per metre):

Conductor operating
temperature: 70 °C

Conductor cross-sectional area	Two-core cable, d.c.	Two-core cable, single-phase a.c.	Three- or four-core cable, three-phase
1	2	3	4
mm²	mV/A/m	mV/A/m	mV/A/m
1	44	44	38
1.5	29	29	25
2.5	18	18	15
4	11	11	9.5
6	7.3	7.3	6.4
10	4.4	4.4	3.8
16	2.8	2.8	2.4
		r	r
25	1.75	1.75	1.50
35	1.25	1.25	1.10
50	0.93	0.93	0.80
70	0.63	0.63	0.55
95	0.46	0.47	0.41

Note: The reactive element of voltage drop usually provided for 25 mm²
and above conductor sizes is omitted for simplicity. For a fuller
treatment see Appendix 4 of BS 7671.

TABLE 6F

Table 4D5A

70 °C thermoplastic (pvc) insulated and sheathed flat cable
with protective conductor

(COPPER CONDUCTORS)
BS 6004 Table 8

CURRENT-CARRYING CAPACITY   Ambient temperature: 30 °C
(amperes):   Conductor operating temperature: 70 °C

Conductor cross-sectional area	Installation Method 6* (Enclosed in conduit in an insulated wall)	Installation Method 15* (Installed directly in an insulated wall)	Reference Method 1 (clipped direct)	Voltage drop: (per ampere per metre)
	1 two-core cable, single-phase a.c. or d.c.			
1	2	3	4	5
(mm²)	(A)	(A)	(A)	(mV/A/m)
1	11.5	12	16	44
1.5	14.5	15	20	29
2.5	20	21	27	18
4	26	27	37	11
6	32	35	47	7.3
10	44	47	64	4.4
16	57	63	85	2.8

Notes:

1   Where the conductor is to be protected by a semi-enclosed fuse to BS 3036, see the introduction to this Appendix.

2   *   These methods are regarded as Reference Methods for the cable types specified by the table.

APPENDIX 7
CERTIFICATION AND REPORTING

The certificates are used with the kind permission of the BSI.

The introduction to Appendix 6 of BS 7671 : 2001 (Model forms for certification and reporting) is reproduced on this page.

Introduction

(i)   The Electrical Installation Certificate required by Part 7 of BS 7671 shall be made out and signed or otherwise authenticated by a competent persons or persons in respect of the design, construction, inspection and testing of the work.

(ii)   The Minor Works Certificate required by Part 7 of BS 7671 shall be made out and signed or otherwise authenticated by a competent person in respect of the inspection and testing of an installation.

(iii)   The Periodic Inspection Report required by Part 7 of BS 7671 shall be made out and signed or otherwise authenticated by a competent persons in respect of the inspection and testing of an installation.

(iv)   Competent persons will, as appropriate to their function under (i) (ii) and (iii) above, have a sound knowledge and experience relevant to the nature of the work undertaken and to the technical standards set down in this British Standard, be fully versed in the inspection and testing procedures contained in this Standard and employ adequate testing equipment.

(v)   Electrical Installation Certificates will indicate the responsibility for design, construction, inspection and testing, whether in relation to new work or further work on an existing installation.

Where design, construction and inspection and testing is the responsibility of one person a Certificate with a single signature declaration in the form shown below may replace the multiple signatures section of the model form.

FOR DESIGN, CONSTRUCTION, INSPECTION & TESTING.

I being the person responsible for the Design, Construction, Inspection & Testing of the electrical installation (as indicated by my signature below), particulars of which are described above, having exercised reasonable skill and care when carrying out the Design, Construction, Inspection & Testing, hereby CERTIFY that the said work for which I have been responsible is to the best of my knowledge and belief in accordance with BS 7671 : ...., amended to ....(date) except for the departures, if any, detailed as follows.

(vi)   A Minor Works Certificate will indicate the responsibility for design, construction, inspection and testing of the work described in Part 4 of the certificate.

(vii)   A Periodic Inspection Report will indicate the responsibility for the inspection and testing of an installation within the extent and limitations specified on the form report.

(viii) A Schedule of Inspections and a Schedule of Test Results as required by Part 7 shall be issued with the associated Electrical Installation Certificate or Periodic Inspection Report.

(ix)   When making out and signing a form on behalf of a company or other business entity, individuals shall state for whom they are acting.

(x)   Additional forms may be required as clarification, if needed by non-technical persons, or in expansion, for larger or more complex installations.

(xi)   The IEE Guidance Note 3 provides further information on inspection and testing on completion and for periodic inspections.

ELECTRICAL INSTALLATION CERTIFICATES
NOTES FOR SHORT FORM F1 AND STANDARD FORM F2:

1.   The Electrical Installation Certificate is to be used only for the initial certification of a new installation or for an alteration or addition to an existing installation where new circuits have been introduced.

It is not to be used for a Periodic Inspection for which a Periodic Inspection Report form should be used. For an alteration or addition which does not extend to the introduction of new circuits, a Minor Electrical Installation Works Certificate may be used.

The original Certificate is to be given to the person ordering the work (Regulation 742-01-03). A duplicate should be retained by the contractor.

2.   This Certificate is only valid if accompanied by the Schedule of Inspections and the Schedule(s) of Test Results.

3.   The signatures appended are those of the persons authorised by the companies executing the work of design, construction and inspection and testing respectively. A signatory authorised to certify more than one category of work should sign in each of the appropriate places.

4.   The time interval recommended before the first periodic inspection must be inserted (see IEE Guidance Note 3 for guidance).

5.   The page numbers for each of the Schedules of Test Results should be indicated, together with the total number of sheets involved.

6.   The maximum prospective fault current recorded should be the greater of either the short-circuit current or the earth fault current.

7.   The proposed date for the next inspection should take into consideration the frequency and quality of maintenance that the installation can reasonably be expected to receive during its intended life, and the period should be agreed between the designer, installer and other relevant parties.

Form F1
ELECTRICAL INSTALLATION CERTIFICATE (notes 1 and 2)   Form No *123*/1

(REQUIREMENTS FOR ELECTRICAL INSTALLATIONS - BS 7671 [IEE WIRING REGULATIONS])

DETAILS OF THE CLIENT (note 1)	*House Builder Ltd, 1, City Way*
	1 City Way, LONDON
INSTALLATION ADDRESS	*Plot 24, New Road*
	NEW TOWN
	County Postcode *AB1 2CD*

DESCRIPTION AND EXTENT OF THE INSTALLATION Tick boxes as appropriate

Description of installation: *Domestic*

Extent of installation covered by this Certificate:

Complete electrical, including
smoke and intruder alarms

(Use continuation sheet if necessary)   see continuation sheet No: ....

New installation	☑
Addition to an existing installation	☐
Alteration to an existing installation	☐

FOR DESIGN, CONSTRUCTION, INSPECTION & TESTING
I being the person responsible for the Design, Construction, Inspection & Testing of the electrical installation (as indicated by my signature below), particulars of which are described above, having exercised reasonable skill and care when carrying out the Design, Construction, Inspection & Testing, hereby CERTIFY that the said work for which I have been responsible is to the best of my knowledge and belief in accordance with BS 7671: ..*2001*.., amended to ...*2002*... (date) except for the departures, if any, detailed as follows:

Details of departures from BS 7671 (Regulations 120-01-03, 120-02):

None

The extent of liability of the signatory is limited to the work described above as the subject of this Certificate.

Name (IN BLOCK LETTERS): *A SMITH*
Signature (note 3): *All Electrics Ltd* *A Smith*   Position: ..*Director*......
For and on behalf of: ....   Date: ....*20.12.2002*...........
Address: ....*27, Central Road*
........*NEW TOWN*
........*County*........ Postcode *EB 4GH*.   Tel No: ....

NEXT INSPECTION
I recommend that this installation is further inspected and tested after an interval of not more than ..*10*...... years/months. (notes 4 and 7)

SUPPLY CHARACTERISTICS AND EARTHING ARRANGEMENTS Tick boxes and enter details, as appropriate

Earthing arrangements	Number and Type of Live Conductors	Nature of Supply Parameters	Supply Protective Device Characteristics
TN-C ☐	a.c. ☑ d.c. ☐	Nominal voltage, $U/U_o^{(1)}$ ...*230*... V	Type: *BS 1361*
TN-S ☐			*fuse*
TN-C-S ☑	1-phase, 2-wire ☑ 2-pole ☐	Nominal frequency, $f^{(1)}$ ....*50*...Hz	
TT ☐	1-phase, 3-wire ☐ 3-pole ☐	Prospective fault current, $I_{pf}^{(2)}$ ..*16*..kA	Nominal current rating
IT ☐		(note 6)	..*100*....A
Alternative source ☐ of supply (to be detailed on attached schedules)	2-phase, 3-wire ☐ other ☐	External loop impedance, $Z_e^{(2)}$ *0.35* Ω	
	3-phase, 3-wire ☐	*(Note: (1) by enquiry, (2) by enquiry or by measurement)*	
	3-phase, 4-wire ☐		

Page 1 of *4* (note 5)

129

PARTICULARS OF INSTALLATION REFERRED TO IN THE CERTIFICATE Tick boxes and enter details, as appropriate

Means of Earthing	Maximum Demand
Supplier's facility ☑	Maximum demand (load) .... *60* ...... Amps per phase

Details of Installation Earth Electrode (*where applicable*)

Installation earth electrode ☐	Type (e.g. rod(s), tape etc) ....*None*....	Location	Electrode resistance to earth .... Ω

Main Protective Conductors

Earthing conductor: material ....*Copper*....... csa ....*1.6*.......mm²   connection verified ☑

Main equipotential bonding conductors   material ....*Copper*....... csa ....*10*.......mm²   connection verified ☑

To incoming water and/or gas service ☑   To other elements: ....

Main Switch or Circuit-breaker

BS, Type ....*BS EN 60439-3*....   No. of poles ....*2*.......   Current rating ....*80*..A   Voltage rating ....*230*..V

Location ....*Garage*...............................   Fuse rating or setting...............A

Rated residual operating current $I_{\Delta n}$ = ...*30*. mA, and operating time of *200*ms (at $I_{\Delta n}$) (applicable only where an RCD is suitable and is used as a main circuit-breaker)

COMMENTS ON EXISTING INSTALLATION (in the case of an alteration or additions see Section 743):

...........*New installation*...

....

....

SCHEDULES (note 2)

The attached Schedules are part of this document and this Certificate is valid only when they are attached to it.
.....*1*...... Inspection Schedules and ....*1*....... Test Result Schedules are attached.
(Enter quantities of schedules attached).

GUIDANCE FOR RECIPIENTS

This safety Certificate has been issued to confirm that the electrical installation work to which it relates has been designed, constructed and inspected and tested in accordance with British Standard 7671 (The IEE Wiring Regulations).

You should have received an original Certificate and the contractor should have retained a duplicate Certificate. If you were the person ordering the work, but not the user of the installation, you should pass this Certificate, or a full copy of it including the schedules, immediately to the user.

The "original" Certificate should be retained in a safe place and be shown to any person inspecting or undertaking further work on the electrical installation in the future. If you later vacate the property, this Certificate will demonstrate to the new owner that the electrical installation complied with the requirements of British Standard 7671 at the time the Certificate was issued. The Construction (Design and Management) Regulations require that for a project covered by those Regulations, a copy of this Certificate, together with schedules is included in the project health and safety documentation.

For safety reasons, the electrical installation will need to be inspected at appropriate intervals by a competent person. The maximum time interval recommended before the next inspection is stated on Page 1 under "Next Inspection".

This Certificate is intended to be issued only for a new electrical installation or for new work associated with an alteration or addition to an existing installation. It should not have been issued for the inspection of an existing electrical installation. A "Periodic Inspection Report" should be issued for such a periodic inspection.

The Certificate is only valid if a Schedule of Inspections and a Schedule of Test Results is appended.

SCHEDULE OF INSPECTIONS

Methods of protection against electric shock

(a) Protection against both direct and indirect contact:

- N/A (i) SELV (note 1)
- N/A (ii) Limitation of discharge of energy

(b) Protection against direct contact: (note 2)

- ✓ (i) Insulation of live parts
- ✓ (ii) Barriers or enclosures
- N/A (iii) Obstacles (note 3)
- N/A (iv) Placing out of reach (note 4)
- N/A (v) PELV
- ✓ (vi) Presence of RCD for supplementary protection

(c) Protection against indirect contact:

- (i) EEBAD including:
- ✓   Presence of earthing conductor
- ✓   Presence of circuit protective conductors
- ✓   Presence of main equipotential bonding conductors
- ✓   Presence of supplementary equipotential bonding conductors
- N/A   Presence of earthing arrangements for combined protective and functional purposes
- N/A   Presence of adequate arrangements for alternative source(s), where applicable
- N/A   Presence of residual current device(s)
- N/A (ii) Use of Class II equipment or equivalent insulation (note 5)
- N/A (iii) Non-conducting location: (note 6) Absence of protective conductors
- N/A (iv) Earth-free equipotential bonding: (note 7) Presence of earth-free equipotential bonding conductors
- N/A (v) Electrical separation (note 8)

Inspected by ..... *A. Smith* ....

Prevention of mutual detrimental influence

- ✓ (a) Proximity of non-electrical services and other influences
- ✓ (b) Segregation of band I and band II circuits or band II insulation used
- ✓ (c) Segregation of safety circuits

Identification

- ✓ (a) Presence of diagrams, instructions, circuit charts and similar information
- ✓ (b) Presence of danger notices and other warning notices
- ✓ (c) Labelling of protective devices, switches and terminals
- ✓ (d) Identification of conductors

Cables and conductors

- ✓ (a) Routing of cables in prescribed zones or within mechanical protection
- ✓ (b) Connection of conductors
- ✓ (c) Erection methods
- ✓ (d) Selection of conductors for current-carrying capacity and voltage drop
- ✓ (e) Presence of fire barriers, suitable seals and protection against thermal effects

General

- ✓ (a) Presence and correct location of appropriate devices for isolation and switching
- ✓ (b) Adequacy of access to switchgear and other equipment
- ✓ (c) Particular protective measures for special installations and locations
- ✓ (d) Connection of single-pole devices for protection or switching in phase conductors only
- ✓ (e) Correct connection of accessories and equipment
- N/A (f) Presence of undervoltage protective devices
- ✓ (g) Choice and setting of protective and monitoring devices for protection against indirect contact and/or overcurrent
- ✓ (h) Selection of equipment and protective measures appropriate to external influences
- ✓ (i) Selection of appropriate functional switching devices

Date ...... *20/12/2002* ....

Notes:

- ✓   to indicate an inspection has been carried out and the result is satisfactory
- ✗   to indicate an inspection has been carried out and the result was unsatisfactory
- N/A   to indicate the inspection is not applicable

1. SELV An extra-low voltage system which is electrically separate from earth and from other systems. The particular requirements of the Regulations must be checked (see Regulations 411-02 and 471-02)

2. Method of protection against direct contact - will include measurement of distances where appropriate

3. Obstacles - only adopted in special circumstances (see Regulations 412-04 and 471-06)

4. Placing out of reach - only adopted in special circumstances (see Regulations 412-05 and 471-07)

5. Use of Class II equipment - infrequently adopted and only when the installation is to be supervised (see Regulations 413-03 and 471-09)

6. Non-conducting locations - not applicable in domestic premises and requiring special precautions (see Regulations 413-04 and 471-10)

7. Earth-free local equipotential bonding - not applicable in domestic premises, only used in special circumstances (see Regulations 413-05 and 471-14)

8. Electrical separation (see Regulations 413-06 and 471-12)

Page *3* of *4*

Form 4
SCHEDULE OF TEST RESULTS

Contractor: ...All Electrics Ltd.........

Test Date: ...20.12.2002...............

Signature ...A. Smith...................

Method of protection against indirect contact: ...E. E. B. A. D. S..................

Equipment vulnerable to testing: ...30 mA RCDs circuits 1 and 4, dimmer and fluorescent circuit 2. Shaver circuit 6.

Address/Location of distribution board:
...Plot 24, New Road..................
...Town..................
...County..................

* Type of Supply: TN-S/TN-C-S/TT

* Ze at origin: 0.35 ohms

* PFC: ...16 kA

Instruments
loop impedance: ...AB11
continuity: ...AB22
insulation: ...AB44
RCD tester: ...AB55

Description of Work: ...House electrical installation...............

Circuit Description	Overcurrent Device * Short-circuit capacity: ...6.kA		Wiring Conductors			Continuity				Insulation Resistance			Test Results					Remarks
	type	Rating I_n	live	cpc		$R_1 + R_2$	R_2	R i n g *8	Live/ Live	Live/ Earth	P o l a r i t y	Earth Loop Imped-ance Z_s	Functional Testing			
													RCD time	Other		
		A	mm²	mm²		Ω	Ω		MΩ	MΩ		Ω	ms			
1	2	3	4	5		*6	*7	*8	*9	*10	*11	*12	*13	*14		15
Lights up	B	10	1.5	1.0		2.4	—	—	50	40	✓	2.8	—			
Lights down	B	10	1.5	1.0		2.7	—	—	—	30	✓	3.1	—	✓		Dimmer
Sockets up	B	32	2.5	1.5		0.4	0.3	✓	30	30	✓	0.8	—	✓		
Sockets down	B	32	2.5	1.5		0.5	0.3	✓	—	30	✓	0.9	200	✓		RCD Vulnerable
Cooker	B	32	6.0	2.5		0.1	—	—	50	40	✓	0.5	—	✓		
Shower	B	45	10.0	4.0		0.15	—	—	—	40	✓	0.5	200	✓		Electronic
Garage	B	20	2.5	1.5		0.4	—	—	—	30	✓	0.8	200	✓		RCD

Deviations from Wiring Regulations and special notes:

None

* See notes on schedule of test results

NOTES ON SCHEDULE OF TEST RESULTS

* **Type of supply** is ascertained from the supply company or by inspection.

* **Z_e at origin.** When the maximum value declared by the electricity supplier is used, the effectiveness of the earth must be confirmed by a test. If measured the main bonding will need to be disconnected for the duration of the test.

* **Short-circuit capacity** of the device is noted, see Table 7.2A of the On-Site Guide or 2.7.15 of GN3

* **Prospective fault current (PFC).** The value recorded is the greater of either the short-circuit current or the earth fault current. Preferably determined by enquiry of the supplier.

The following tests, where relevant, shall be carried out in the following sequence:

Continuity of protective conductors, including main and supplementary bonding
Every protective conductor, including main and supplementary bonding conductors, should be tested to verify that they are continuous and correctly connected.

*6 **Continuity**
Where Test Method 1 is used, enter the measured resistance of the phase conductor plus the circuit protective conductor ($R_1 + R_2$).
See 10.3.1 of the On-Site Guide or 2.7.5 of GN3.
During the continuity testing (Test Method 1) the following polarity checks are to be carried out:
(a)   every fuse and single-pole control and protective device is connected in the phase conductor only
(b)   centre-contact bayonet and Edison screw lampholders have outer contact connected to the neutral conductor
(c)   wiring is correctly connected to socket-outlets and similar accessories.
Compliance is to be indicated by a tick in polarity column 11.

*7   Where Test Method 2 is used, the maximum value of R_2 is recorded in column 7.
Where the alternative method of Regulation 413-02-12 is used for shock protection, the resistance of the circuit protective conductor R_2 is measured and recorded in column 7.
See 10.3.1 of the On-Site Guide or 2.7.5 of GN3.

*8 **Continuity of ring final conductors**
A test shall be made to verify the continuity of each conductor including the protective conductor of every ring final circuit.
See 10.3.2 of the On-Site Guide or 2.7.6 of GN3.

*9, *10 **Insulation Resistance**
All voltage sensitive devices to be disconnected or test between live conductors (phase and neutral) connected together and earth.
The insulation resistance between live conductors is to be inserted in column 9.
The minimum insulation resistance values are given in Table 10.1 of the On-Site Guide or Table 2.2 of GN3.
See 10.3.3(iv) of the On-Site Guide or 2.7.7 of GN3.

All the preceding tests should be carried out before the installation is energised.

*11 **Polarity**
A satisfactory polarity test may be indicated by a tick in column 11.
Only in a Schedule of Test Results associated with a Periodic Inspection Report is it acceptable to record incorrect polarity.

*12 **Earth fault loop impedance Z_s**
This may be determined either by direct measurement at the furthest point of a live circuit or by adding ($R_1 + R_2$) of column 6 to Z_e. Z_e is determined by measurement at the origin of the installation or preferably the value declared by the supply company used.
$Z_s = Z_e + (R_1 + R_2)$. Z_s should be less than the values given in Appendix 2 of the On-Site Guide or App 2 of GN3.

*13 **Functional testing**
The operation of RCDs (including RCBOs) shall be tested by simulating a fault condition, independent of any test facility in the device.
Record operating time in column 13. Effectiveness of the test button must be confirmed.
See Section 11 of the On-Site Guide or 2.7.16 of GN3

*14 All switchgear and controlgear assemblies, drives, control and interlocks, etc must be operated to ensure that they are properly mounted, adjusted, and installed.
Satisfactory operation is indicated by a tick in column 14.

Earth electrode resistance
The earth electrode resistance of TT installations must be measured, and normally an RCD is required.
For reliability in service the resistance of any earth electrode should be below 200 Ω. Record the value on Forms 1, 2 or 6 as appropriate. See 10.3.5 of the On-Site Guide or 2.7.13 of GN3.

ELECTRICAL INSTALLATION CERTIFICATE (notes 1 and 2)
(REQUIREMENTS FOR ELECTRICAL INSTALLATIONS - BS 7671 [IEE WIRING REGULATIONS])

DETAILS OF THE CLIENT (note 1)	*Adeveloper Ltd*
	Main Street
	London, E11

INSTALLATION ADDRESS	*Plot 13, Industrial Site*
	Town
	County .... Postcode *CB1 2TT*

DESCRIPTION AND EXTENT OF THE INSTALLATION Tick boxes as appropriate
(note 1)

Description of installation: *Warehouse/factory with office*

	New installation	☑
Extent of installation covered by this Certificate: *Complete installation*	Addition to an existing installation	☐
	Alteration to an existing installation	☐

FOR DESIGN
I/We being the person(s) responsible for the design of the electrical installation (as indicated by my/our signatures below), particulars of which are described above, having exercised reasonable skill and care when carrying out the design hereby CERTIFY that the design work for which I/we have been responsible is to the best of my/our knowledge and belief in accordance with BS 7671 : *2001*., amended to .... *2002* ....(date) except for the departures, if any, detailed as follows:

> Details of departures from BS 7671 (Regulations 120-01-03, 120-02):
> *None*

The extent of liability of the signatory or the signatories is limited to the work described above as the subject of this Certificate.

For the DESIGN of the installation: **(Where there is mutual responsibility for the design)

Signature: *B Brown* .... Date: *5/2/2002* ...... Name (BLOCK LETTERS): ...... *B BROWN* .... Designer No 1

Signature:.................... Date:............ Name (BLOCK LETTERS): .... Designer No 2**

FOR CONSTRUCTION
I/We being the person(s) responsible for the construction of the electrical installation (as indicated by my/our signatures below), particulars of which are described above, having exercised reasonable skill and care when carrying out the construction hereby CERTIFY that the construction work for which I/we have been responsible is to the best of my/our knowledge and belief in accordance with BS 7671 : *2001*., amended to .... *2002* ....(date) except for the departures, if any, detailed as follows:

> Details of departures from BS 7671 (Regulations 120-01-03, 120-02):
> *None*

The extent of liability of the signatory is limited to the work described above as the subject of this Certificate.

For CONSTRUCTION of the installation:
Signature *W White* .... Date *6/2/2002*
Name (BLOCK LETTERS): ..... *W WHITE* .... Constructor

FOR INSPECTION & TESTING
I/We being the person(s) responsible for the inspection & testing of the electrical installation (as indicated by my/our signatures below), particulars of which are described above, having exercised reasonable skill and care when carrying out the inspection & testing hereby CERTIFY that the work for which I/we have been responsible is to the best of my/our knowledge and belief in accordance with BS 7671 : *2001*., amended to .... *2002* ....(date) except for the departures, if any, detailed as follows:

> Details of departures from BS 7671 (Regulations 120-01-03, 120-02):
> *None*

The extent of liability of the signatory is limited to the work described above as the subject of this Certificate.

For INSPECTION AND TEST of the installation:
Signature:..... *S Jones* .... Date: .. *10/2/2002* ....
Name (BLOCK LETTERS): ..... *S JONES* .... Inspector

NEXT INSPECTION (notes 4 and 7)
I/We the designer(s), recommend that this installation is further inspected and tested after an interval of not more than ..... *3* .... years/~~months~~.

Page 1 of *4*

134

PARTICULARS OF SIGNATORIES TO THE ELECTRICAL INSTALLATION CERTIFICATE

Designer (No 1)
Name: *B Brown*   Company: *Design Co Ltd*
Address: *City Road, Old Town*
County   Postcode: *MP1 2BQ* Tel No: *01334/612266*

Designer (No 2)
(if applicable)
Name: *Not applicable*   Company:
Address:
Postcode: Tel No:

Constructor
Name: *W White*   Company: *County Electrics Ltd*
Address: *18, Industrial Lane, Town*
County   Postcode: *MP3 8BQ* Tel No: *01334/72963*

Inspector
Name: *S Jones*   Company: *County Electrics Ltd*
Address: *As above*
Postcode: Tel No:

SUPPLY CHARACTERISTICS AND EARTHING ARRANGEMENTS Tick boxes and enter details, as appropriate

Earthing arrangements	Number and Type of Live Conductors	Nature of Supply Parameters	Supply Protective Device Characteristics
TN-C ☐ TN-S ☐ TN-C-S ☑ TT ☐ IT ☐	a.c. ☑ d.c. ☐ 1-phase, 2-wire ☐ 2-pole ☐ 1-phase, 3-wire ☐ 3-pole ☐ 2-phase, 3-wire ☐ other ☐ 3-phase, 3-wire ☐ 3-phase, 4-wire ☑	Nominal voltage, U/U₀ $^{(1)}$ *400/230* V Nominal frequency, f $^{(1)}$ *50* Hz (note 6) Prospective fault current, I$_{pf}$ $^{(2)}$ *18* kA External loop impedance, Z$_e$ $^{(2)}$ *0.2* Ω (Note: (1) by enquiry, (2) by enquiry or by measurement)	Type: *BS 1361* *Fuse* Nominal current rating *100* A

Alternative source ☐
of supply (to be detailed
on attached schedules)

PARTICULARS OF INSTALLATION REFERRED TO IN THE CERTIFICATE Tick boxes and enter details, as appropriate

Means of Earthing

Supplier's facility ☑

Installation ☐
earth electrode

Maximum Demand

Maximum demand (load) *40* Amps per phase

Details of Installation Earth Electrode (*where applicable*)

Type   Location   Electrode resistance to earth
(e.g. rod(s), tape etc)
.... Ω

Main Protective Conductors

Earthing conductor: material *copper* csa *16* mm²   connection verified ☑

Main equipotential bonding
conductors   material *copper* csa *10* mm²   connection verified ☑

To incoming water and/or gas service ☑   To other elements:

Main Switch or Circuit-breaker

BS, Type *BS EN 60947-3* No. of poles *3* Current rating *125* A   Voltage rating *400* V

Location *Switchroom adjacent main office*   Fuse rating or setting............A

Rated residual operating current I$_{\Delta n}$ = .../..... mA, and operating time of ...../..... ms (at I$_{\Delta n}$) (applicable only where an RCD is suitable and is used as a main circuit-breaker)

COMMENTS ON EXISTING INSTALLATION (in the case of an alteration or additions see Section 743):
Not applicable

SCHEDULES (note 2)
The attached Schedules are part of this document and this Certificate is valid only when they are attached to it.
1 Inspection Schedules and *1* Test Result Schedules are attached.
(Enter quantities of schedules attached).

ELECTRICAL INSTALLATION CERTIFICATE
GUIDANCE FOR RECIPIENTS (to be appended to the Certificate)

This safety Certificate has been issued to confirm that the electrical installation work to which it relates has been designed, constructed and inspected and tested in accordance with British Standard 7671 (The IEE Wiring Regulations).

You should have received an original Certificate and the contractor should have retained a duplicate Certificate. If you were the person ordering the work, but not the user of the installation, you should pass this Certificate, or a full copy of it including the schedules, immediately to the user.

The "original" Certificate should be retained in a safe place and be shown to any person inspecting or undertaking further work on the electrical installation in the future. If you later vacate the property, this Certificate will demonstrate to the new owner that the electrical installation complied with the requirements of British Standard 7671 at the time the Certificate was issued. The Construction (Design and Management) Regulations require that for a project covered by those Regulations, a copy of this Certificate, together with schedules is included in the project health and safety documentation.

For safety reasons, the electrical installation will need to be inspected at appropriate intervals by a competent person. The maximum time interval recommended before the next inspection is stated on Page 1 under "Next Inspection".

This Certificate is intended to be issued only for a new electrical installation or for new work associated with an alteration or addition to an existing installation. It should not have been issued for the inspection of an existing electrical installation. A "Periodic Inspection Report" should be issued for such a periodic inspection.

The Certificate is only valid if a Schedule of Inspections and a Schedule of Test Results is appended.

SCHEDULE OF INSPECTIONS

Methods of protection against electric shock

(a) Protection against both direct and indirect contact:

N/A	(i)	SELV (note 1)
N/A	(ii)	Limitation of discharge of energy

(b) Protection against direct contact: (note 2)

✓	(i)	Insulation of live parts
✓	(ii)	Barriers or enclosures
N/A	(iii)	Obstacles (note 3)
N/A	(iv)	Placing out of reach (note 4)
N/A	(v)	PELV
✓	(vi)	Presence of RCD for supplementary protection

(c) Protection against indirect contact:

	(i)	EEBAD including:
✓		Presence of earthing conductor
✓		Presence of circuit protective conductors
✓		Presence of main equipotential bonding conductors
✓		Presence of supplementary equipotential bonding conductors
N/A		Presence of earthing arrangements for combined protective and functional purposes
N/A		Presence of adequate arrangements for alternative source(s), where applicable
N/A		Presence of residual current device(s)
N/A	(ii)	Use of Class II equipment or equivalent insulation (note 5)
N/A	(iii)	Non-conducting location: (note 6) Absence of protective conductors
N/A	(iv)	Earth-free equipotential bonding: (note 7) Presence of earth-free equipotential bonding conductors
N/A	(v)	Electrical separation (note 8)

Inspected by *S. Jones* .... Date *10/2/2002* ....

Prevention of mutual detrimental influence

✓	(a)	Proximity of non-electrical services and other influences
✓	(b)	Segregation of band I and band II circuits or band II insulation used
✓	(c)	Segregation of safety circuits

Identification

✓	(a)	Presence of diagrams, instructions, circuit charts and similar information
✓	(b)	Presence of danger notices and other warning notices
✓	(c)	Labelling of protective devices, switches and terminals
✓	(d)	Identification of conductors

Cables and conductors

✓	(a)	Routing of cables in prescribed zones or within mechanical protection
✓	(b)	Connection of conductors
✓	(c)	Erection methods
✓	(d)	Selection of conductors for current-carrying capacity and voltage drop
✓	(e)	Presence of fire barriers, suitable seals and protection against thermal effects

General

✓	(a)	Presence and correct location of appropriate devices for isolation and switching
✓	(b)	Adequacy of access to switchgear and other equipment
✓	(c)	Particular protective measures for special installations and locations
✓	(d)	Connection of single-pole devices for protection or switching in phase conductors only
✓	(e)	Correct connection of accessories and equipment
N/A	(f)	Presence of undervoltage protective devices
✓	(g)	Choice and setting of protective and monitoring devices for protection against indirect contact and/or overcurrent
✓	(h)	Selection of equipment and protective measures appropriate to external influences
✓	(i)	Selection of appropriate functional switching devices

Notes:

✓   to indicate an inspection has been carried out and the result is satisfactory
X   to indicate an inspection has been carried out and the result was unsatisfactory
N/A   to indicate the inspection is not applicable

SELV An extra-low voltage system which is electrically separate from earth and from other systems. The particular requirements of the Regulations must be checked (see Regulations 411-02 and 471-02)

Method of protection against direct contact - will include measurement of distances where appropriate

Obstacles - only adopted in special circumstances (see Regulations 412-04 and 471-06)

Placing out of reach - only adopted in special circumstances (see Regulations 412-05 and 471-07)

5.   Use of Class II equipment - infrequently adopted and only when the installation is to be supervised (see Regulations 413-03 and 471-09)

6.   Non-conducting locations - not applicable in domestic premises and requiring special precautions (see Regulations 413-04 and 471-10)

7.   Earth-free local equipotential bonding - not applicable in domestic premises, only used in special circumstances (see Regulations 413-05 and 471-14)

8.   Electrical separation (see Regulations 413-06 and 471-12)

Page *3* **of** *4*

Form 4
SCHEDULE OF TEST RESULTS

Contractor: Elec.Contracts Ltd.
Test Date: 10.12.2002
Signature: S. Jones
Method of protection against indirect contact: E.E.B.A.D.S.
Equipment vulnerable to testing: Luminaire controllers, circuits 1, 2 and 3.

Address/Location of distribution board:
Plot 13, Industrial Site

* Type of Supply: ~~TN-S~~/TN-C-S/~~TT~~
* Ze at origin: 0.2 ohms
* PFC: 18 kA

Instruments
loop impedance: AB11
continuity: AB27
insulation: AB44
RCD tester: AB65

Description of Work: Warehouse and office

Circuit Description	Overcurrent Device		Wiring Conductors		Continuity			Insulation Resistance		Polarity	Earth Loop Imped-ance	Functional Testing		Remarks
	type	Rating I_n	live	cpc	$R_1 + R_2$	R_2	R ring	Live/Live	Live/Earth		Z_s	RCD time	Other	
		A	mm²	mm²	Ω	Ω	n g	MΩ	MΩ		Ω	ms		
1	2	3	4	5	*6	*7	*8	*9	*10	*11	*12	*13	*14	15
Lights R	BS 88	16	2.5	1.5	2.0	—	—	—	10	✓	2.2		✓	Vulnerable
Lights Y	BS 88	16	2.5	1.5	2.3	—	—	—	10	✓	2.5		✓	Vulnerable
Lights B	BS 88	16	2.5	1.5	1.6	—	\	\	10	✓	1.8		✓	Vulnerable
Sockets R	BS 88	32	2.5	1.5	0.5	0.3	✓	30	20	✓	0.7		✓	
Sockets Y	BS 88	32	2.5	1.5	0.4	0.3	—	30	20	✓	0.6		✓	
Bus Bar R	BS 88	63	16	10	0.1	—	—	40	30	✓	0.32		✓	
Bus Bar Y	BS 88	63	16	10	0.1	—	—	40	30	✓	0.32		✓	
Bus Bar B	BS 88	63	16	10	0.1	—	—	40	30	✓	0.32		✓	

Test Results

Deviations from Wiring Regulations and special notes:

None

* See notes on schedule of test results

Page 4 of 4

138

NOTES ON SCHEDULE OF TEST RESULTS

* **Type of supply** is ascertained from the supply company or by inspection.

* **Z_e at origin.** When the maximum value declared by the electricity supplier is used, the effectiveness of the earth must be confirmed by a test. If measured the main bonding will need to be disconnected for the duration of the test.

* **Short-circuit capacity** of the device is noted, see Table 7.2A of the On-Site Guide or 2.7.15 of GN3

* **Prospective fault current (PFC).** The value recorded is the greater of either the short-circuit current or the earth fault current. Preferably determined by enquiry of the supplier.

The following tests, where relevant, shall be carried out in the following sequence:

Continuity of protective conductors, including main and supplementary bonding

Every protective conductor, including main and supplementary bonding conductors, should be tested to verify that they are continuous and correctly connected.

*6 **Continuity**

Where Test Method 1 is used, enter the measured resistance of the phase conductor plus the circuit protective conductor ($R_1 + R_2$). See 10.3.1 of the On-Site Guide or 2.7.5 of GN3.
During the continuity testing (Test Method 1) the following polarity checks are to be carried out:
(a)   every fuse and single-pole control and protective device is connected in the phase conductor only
(b)   centre-contact bayonet and Edison screw lampholders have outer contact connected to the neutral conductor
(c)   wiring is correctly connected to socket-outlets and similar accessories.
Compliance is to be indicated by a tick in polarity column 11.

*7   Where Test Method 2 is used, the maximum value of R_2 is recorded in column 7.
Where the alternative method of Regulation 413-02-12 is used for shock protection, the resistance of the circuit protective conductor R_2 is measured and recorded in column 7.
See 10.3.1 of the On-Site Guide or 2.7.5 of GN3.

*8 **Continuity of ring final conductors**

A test shall be made to verify the continuity of each conductor including the protective conductor of every ring final circuit. See 10.3.2 of the On-Site Guide or 2.7.6 of GN3.

*9, *10 **Insulation Resistance**

All voltage sensitive devices to be disconnected or test between live conductors (phase and neutral) connected together and earth. The insulation resistance between live conductors is to be inserted in column 9.
The minimum insulation resistance values are given in Table 10.1 of the On-Site Guide or Table 2.2 of GN3.
See 10.3.3(iv) of the On-Site Guide or 2.7.7 of GN3.

All the preceding tests should be carried out before the installation is energised.

*11 **Polarity**

A satisfactory polarity test may be indicated by a tick in column 11.
Only in a Schedule of Test Results associated with a Periodic Inspection Report is it acceptable to record incorrect polarity.

*12 **Earth fault loop impedance Z_S**

This may be determined either by direct measurement at the furthest point of a live circuit or by adding ($R_1 + R_2$) of column 6 to Z_e. Z_e is determined by measurement at the origin of the installation or preferably the value declared by the supply company used. $Z_S = Z_e + (R_1 + R_2)$. Z_S should be less than the values given in Appendix 2 of the On-Site Guide or App 2 of GN3.

*13 **Functional testing**

The operation of RCDs (including RCBOs) shall be tested by simulating a fault condition, independent of any test facility in the device. Record operating time in column 13. Effectiveness of the test button must be confirmed.
See Section 11 of the On-Site Guide or 2.7.16 of GN3

*14 All switchgear and controlgear assemblies, drives, control and interlocks, etc must be operated to ensure that they are properly mounted, adjusted, and installed.
Satisfactory operation is indicated by a tick in column 14.

Earth electrode resistance

The earth electrode resistance of TT installations must be measured, and normally an RCD is required.
For reliability in service the resistance of any earth electrode should be below 200 Ω. Record the value on Forms 1, 2 or 6 as appropriate. See 10.3.5 of the On-Site Guide or 2.7.13 of GN3.

NOTES ON COMPLETION OF MINOR ELECTRICAL INSTALLATION WORKS CERTIFICATE

Scope

The Minor Electrical Installation Works Certificate form is only to be used for additions to an electrical installation that do not extend to the introduction of a new circuit e.g. the addition of a socket-outlet or a lighting point to an existing circuit (Regulation 741-01-03).

Part 1 Description of minor works

1,2   The minor works must be so described that the work that is the subject of the certification can be readily identified.

4   See Regulations 120-05-01, 120-04-01, 120-01-02. No departures are to be expected except in most unusual circumstances. See also Regulation 743-01-03.

Part 2 Installation details

2   The method of protection against indirect contact shock must be clearly identified e.g. earthed equipotential bonding and automatic disconnection of supply using fuse/circuit breaker/RCD

4   If the existing installation lacks either an effective means of earthing or adequate main equipotential bonding conductors, this must be clearly stated. See Regulation 743-01-02.

      Recorded departures from BS 7671 may constitute non-compliance with the Electricity Supply Regulations 1988 as amended or the Electricity at Work Regulations 1989. It is important that the client is advised immediately in writing.

Part 3 Essential Tests

The relevant provisions of Part 7 (Inspection and Testing) of the BS 7671 must be applied in full to all minor works. For example where a socket-outlet is added to an existing circuit it is necessary to:

1   establish that the earthing contact of the socket-outlet is connected to the main earthing terminal

2   measure the insulation resistance of the circuit that has been added to, and establish that it complies with Table 71A of BS 7671

3   measure the earth fault impedance to establish that the maximum permitted disconnection time is not exceeded

4   check that the polarity of e.g. the socket-outlet, is correct

5   (if the work is protected by an RCD) verify the effectiveness of the RCD.

Part 4 Declaration

1,3   The Certificate shall be made out and signed by a competent person in respect of the design, construction, inspection and testing of the work.

1,3   The competent person will have a sound knowledge and experience relevant to the nature of the installation undertaken and to the technical standards set down in BS 7671, be fully versed in the inspection and testing procedures contained in the Regulations and employ adequate testing equipment.

2   When making out and signing a form on behalf of a company other business entity, individuals shall state for whom they are acting.

MINOR ELECTRICAL INSTALLATION WORKS CERTIFICATE

(REQUIREMENTS FOR ELECTRICAL INSTALLATIONS - BS 7671 [IEE WIRING REGULATIONS])
To be used only for minor electrical work which does not include the provision of a new circuit

---

PART 1 : Description of minor works

1. Description of the minor works : .... *Additional socket in main office* ....

2. Location/Address : .... *Office Co Ltd, 20 New Town Road* ....

3. Date minor works completed : .... *21 /10 /2002* ....

4. Details of departures, if any, from BS 7671
   ....
   .... *None* ....
   ....
   ....

PART 2 : Installation details

1. System earthing arrangement   TN-C-S ☑   TN-S ☐   TT ☐

2. Method of protection against indirect contact: ...*E.E.B.A.D.S.*...........................

3. Protective device for the modified circuit   :   Type BS ...*1361*.................   Rating ..*30*........ A

4. Comments on existing installation, including adequacy of earthing and bonding arrangements : (See Regulation 130-07)
   ....
   ....
   .... *None* ....

PART 3 : Essential Tests

1. Earth continuity : satisfactory   ☑

2. Insulation resistance:
   Phase/neutral ....*20*............. MΩ

   Phase/earth ....*20*............. MΩ

   Neutral/earth ....*20*............. MΩ

3. Earth fault loop impedance ....*0.8*.................. Ω

4. Polarity : satisfactory   ☑

5. RCD operation (if applicable) : Rated residual operating current I$_{\Delta n}$ .../.......mA and operating time of .../.......ms (at I$_{\Delta n}$)

PART 4 : Declaration

1. I/We CERTIFY that the said works do not impair the safety of the existing installation, that the said works have been designed, constructed, inspected and tested in accordance with BS 7671 : .*2001*. (IEE Wiring Regulations), amended to ....*2002*........... and that the said works, to the best of my/our knowledge and belief, at the time of my/our inspection, complied with BS 7671 except as detailed in Part 1.

2. Name: ..*W. WHITE*........................   3. Signature:*W. White*...........................

   For and on behalf of: .*County Electrics*...............   Position: ...*Electrician*...........................

   Address: .....*187 Industrial Lane*.....................
   ....   Date: ....*10/2 /2002*................................
   ....

141

MINOR ELECTRICAL INSTALLATION WORKS CERTIFICATE
GUIDANCE FOR RECIPIENTS (to be appended to the Certificate)

This Certificate has been issued to confirm that the electrical installation work to which it relates has been designed, constructed and inspected and tested in accordance with British Standard 7671, (The IEE Wiring Regulations.)

You should have received an original Certificate and the contractor should have retained a duplicate. If you were the person ordering the work, but not the owner of the installation, you should pass this Certificate, or a copy of it, to the owner.

The Minor Works Certificate is only to be used for additions, alterations or replacements to an installation that do not extend to the provision of a new circuit. Examples include the addition of a socket-outlet or lighting point to an existing circuit, or the replacement or relocation of a light switch. A separate Certificate should have been received for each existing circuit on which minor works have been carried out. This Certificate is not valid if you requested the contractor to undertake more extensive installation work. An Electrical Installation Certificate would be required in such circumstances.

The "original" Certificate should be retained in a safe place and be shown to any person inspecting or undertaking further work on the electrical installation in the future. If you later vacate the property, this Certificate will demonstrate to the new owner that the minor electrical installation work carried out complied with the requirements of British Standard 7671 at the time the Certificate was issued.

PERIODIC INSPECTION REPORT
NOTES:

1. This Periodic Inspection Report form shall only be used for the reporting on the condition of an existing installation.

2. The Report, normally comprising at least four pages, shall include schedules of both the inspection and the test results. Additional sheets of test results may be necessary for other than a simple installation. The page numbers of each sheet shall be indicated, together with the total number of sheets involved. The Report is only valid if a Schedule of Inspections and a Schedule of Test Results are appended.

3. The intended purpose of the Periodic Inspection Report shall be identified, together with the recipient's details in the appropriate boxes.

4. The maximum prospective fault current recorded should be the greater of either the short-circuit current or the earth fault current.

5. The 'Extent and Limitations' box shall fully identify the elements of the installation that are covered by the report and those that are not; this aspect having been agreed with the client and other interested parties before the inspection and testing is carried out.

6. The recommendation(s), if any, shall be categorised using the numbered coding 1-4 as appropriate.

7. The 'Summary of the Inspection' box shall clearly identify the condition of the installation in terms of safety.

8. Where the periodic inspection and testing has resulted in a satisfactory overall assessment, the time interval for the next periodic inspection and testing shall be given. The IEE Guidance Note 3 provides guidance on the maximum interval between inspections for various types of buildings. If the inspection and testing reveal that parts of the installation require urgent attention, it would be appropriate to state an earlier re-inspection date having due regard to the degree of urgency and extent of the necessary remedial work.

9. If the space available on the model form for information on recommendations is insufficient, additional pages shall be provided as necessary.

Form F 6   Form No *126*/6

PERIODIC INSPECTION REPORT FOR AN ELECTRICAL INSTALLATION (note 1)

(REQUIREMENTS FOR ELECTRICAL INSTALLATIONS - BS 7671 [IEE WIRING REGULATIONS])

DETAILS OF THE CLIENT

Client: ....*Mr. A. Brown*...

Address: .....*111. Any Street. Town. County. NT7. 8BS*..

Purpose for which this Report is required: .....*Mortgage*...(note 3)

DETAILS OF THE INSTALLATION Tick boxes as appropriate

Occupier: ....*As above*.................

Installation: ...*As above*................

Address: .....*As above*...................

Description of Premises: Domestic ☑   Commercial ☐   Industrial ☐   Other ☐
.....*House with garage*...

Estimated age of the Electrical   ....*15*...... years
Installation:

Evidence of Alterations or Additions: Yes ☑   No ☐   Not apparent ☐

If "Yes", estimate age: ....*5*............ years

Date of last inspection: ....——.................   Records available   Yes ☐   No ☑

EXTENT AND LIMITATIONS OF THE INSPECTION (note 5)

Extent of electrical installation covered by this report: ......*installation to house, garage and garden shed*...........
....
....

Limitations: ....*No dismantling or lifting of floorboards*..
....
....

This inspection has been carried out in accordance with BS 7671: 2001 (IEE Wiring Regulations), amended to *2002* ...
Cables concealed within trunking and conduits, or cables and conduits concealed under floors, in roof spaces and
generally within the fabric of the building or underground have not been inspected.

NEXT INSPECTION (note 8)

I/We recommend that this installation is further inspected and tested after an interval of not more than ...*10*...... months/years,
provided that any observations 'requiring urgent attention' are attended to without delay.

DECLARATION

INSPECTED AND TESTED BY

Name: .....*W. White*..........................   Signature: ......*W White*....................................

For and on behalf of: ......*County Electrics*...........   Position: ......*Electrician*...................................

Address: ...*187 Industrial Lane*..................
..............*Town*.........................   Date: ....*18/4/2002*..
..............*County. MP3 8BQ*........................

SUPPLY CHARACTERISTICS AND EARTHING ARRANGEMENTS _{Tick boxes and enter details, as appropriate}

Earthing arrangements	Number and Type of Live Conductors	Nature of Supply Parameters	Supply Protective Device Characteristics
TN-C ☐ TN-S ☐ TN-C-S ☑ TT ☐ IT ☐	a.c. ☐ ☑ d.c. ☐ 1-phase, 2-wire ☑ 2-pole ☐ 1-phase, 3 wire ☐ 3-pole ☐ 2-phase, 3-wire ☐ other ☐	Nominal voltage, U/U_0 [1] ....230.... V Nominal frequency, f [1] ....50.....Hz Prospective fault current, I_{pf} [2] .....1.0..kA (note 4) External loop impedance, Z_e [2] ...0.24Ω	Type:..BS.1361.... .....fuse...... Nominal current rating ....100.....A
Alternative source ☐ of supply (to be detailed on attached schedules)	3-phase, 3-wire ☐ 3-phase, 4-wire ☐	(Note: (1) by enquiry, (2) by enquiry or by measurement)	

PARTICULARS OF INSTALLATION REFERRED TO IN THE REPORT _{Tick boxes and enter details, as appropriate}

Means of Earthing
Supplier's facility ☑
Installation
earth electrode

Details of Installation Earth Electrode (where applicable)

Type (e.g. rod(s), tape etc)	Location	Electrode resistance to earth
....	....	.... Ω

Main Protective Conductors

Earthing conductor: material ....Copper.... csa ......10....mm² connection verified ☑
Main equipotential bonding conductors material ....Copper.... csa ....6.....mm² connection verified ☑

To incoming water service ☑ To incoming gas service ☑ To incoming oil service ☐ To structural steel ☐
To lightning protection ☐ To other incoming service(s) ☐ (state details...)

Main Switch or Circuit-breaker

BS, Type ..5486........... No. of poles ...2...... Current rating ..80.....A Voltage rating ....240..V
Location....Meter.cupboard............................ Fuse rating or setting ....A

Rated residual operating current $I_{\Delta n}$ = ... mA, and operating time of ..... ms (at $I_{\Delta n}$) _(applicable only where an RCD is suitable and is used as a main circuit-breaker)

OBSERVATIONS AND RECOMMENDATIONS _{Tick boxes as appropriate}

(note 9)
Referring to the attached Schedule(s) of Inspection and Test Results, and subject to the limitations specified at the Extent and Limitations of the Inspection section

☐ No remedial work is required   ☑ The following observations are made:

Recommendations as detailed below note 6

Main bonding conductors sized per 15 edition	4
No 30 mA RCD to socket in garage	1
No 30 mA RCD to socket in shed	1
Light pendants and lampholders worn/faulty	1
No supplementary bonding in bathroom	2
Broken socket in lounge	1

One of the following numbers, as appropriate, is to be allocated to each of the observations made above to indicate to the person(s) responsible for the installation the action recommended.

| 1 | requires urgent attention | 2 | requires improvement | 3 | requires further investigation |

| 4 | does not comply with BS 7671: 2001 amended to ..2002. This does not imply that the electrical installation inspected is unsafe.

SUMMARY OF THE INSPECTION (note 7)

Date(s) of the inspection: ......28/6/2002..........
General condition of the installation: ......Some urgent repairs and improvements required..................
....
....
....
Overall assessment: ~~Satisfactory~~/Unsatisfactory (note 8)

SCHEDULE(S)

The attached Schedules are part of this document and this Report is only valid when they are attached to it.
...1.... Inspection Schedules and ...1.... Test Result Schedules are attached.
_{(Enter quantities of schedules attached).}

PERIODIC INSPECTION REPORT
GUIDANCE FOR RECIPIENTS (to be appended to the Report)

This Periodic Inspection Report form is intended for reporting on the condition of an existing electrical installation.

You should have received an original Report and the contractor should have retained a duplicate. If you were the person ordering this Report, but not the owner of the installation, you should pass this Report, or a copy of it, immediately to the owner.

The original Report is to be retained in a safe place and be shown to any person inspecting or undertaking work on the electrical installation in the future. If you later vacate the property, this Report will provide the new owner with details of the condition of the electrical installation at the time the Report was issued.

The 'Extent and Limitations' box should fully identify the extent of the installation covered by this Report and any limitations on the inspection and tests. The contractor should have agreed these aspects with you and with any other interested parties (Licensing Authority, Insurance Company, Building Society etc) before the inspection was carried out.

The Report will usually contain a list of recommended actions necessary to bring the installation up to the current standard. **For items classified as 'requires urgent attention', the safety of those using the installation may be at risk,** and it is recommended that a competent person undertakes the necessary remedial work without delay.

For safety reasons, the electrical installation will need to be re-inspected at appropriate intervals by a competent person. The maximum time interval recommended before the next inspection is stated in the Report under 'Next Inspection.'

The Report is only valid if a Schedule of Inspections and a Schedule of Test Results is appended.

146

SCHEDULE OF INSPECTIONS

Methods of protection against electric shock

(a) Protection against both direct and indirect contact:

N/A	(i)	SELV (note 1)
N/A	(ii)	Limitation of discharge of energy

(b) Protection against direct contact: (note 2)

✓	(i)	Insulation of live parts
✗	(ii)	Barriers or enclosures
N/A	(iii)	Obstacles (note 3)
N/A	(iv)	Placing out of reach (note 4)
N/A	(v)	PELV
✗	(vi)	Presence of RCD for supplementary protection

(c) Protection against indirect contact:

	(i)	EEBAD including:
✓		Presence of earthing conductor
✓		Presence of circuit protective conductors
✓		Presence of main equipotential bonding conductors
✗		Presence of supplementary equipotential bonding conductors
N/A		Presence of earthing arrangements for combined protective and functional purposes
N/A		Presence of adequate arrangements for alternative source(s), where applicable
N/A		Presence of residual current device(s)
N/A	(ii)	Use of Class II equipment or equivalent insulation (note 5)
N/A	(iii)	Non-conducting location: (note 6) Absence of protective conductors
N/A	(iv)	Earth-free equipotential bonding: (note 7) Presence of earth-free equipotential bonding conductors
N/A	(v)	Electrical separation (note 8)

Prevention of mutual detrimental influence

✓	(a)	Proximity of non-electrical services and other influences
✓	(b)	Segregation of band I and band II circuits or band II insulation used
✓	(c)	Segregation of safety circuits

Identification

✓	(a)	Presence of diagrams, instructions, circuit charts and similar information
✓	(b)	Presence of danger notices and other warning notices
✓	(c)	Labelling of protective devices, switches and terminals
✓	(d)	Identification of conductors

Cables and conductors

✓	(a)	Routing of cables in prescribed zones or within mechanical protection
✓	(b)	Connection of conductors
✓	(c)	Erection methods
✓	(d)	Selection of conductors for current-carrying capacity and voltage drop
✓	(e)	Presence of fire barriers, suitable seals and protection against thermal effects

General

✓	(a)	Presence and correct location of appropriate devices for isolation and switching
✓	(b)	Adequacy of access to switchgear and other equipment
✗	(c)	Particular protective measures for special installations and locations
✓	(d)	Connection of single-pole devices for protection or switching in phase conductors only
✓	(e)	Correct connection of accessories and equipment
N/A	(f)	Presence of undervoltage protective devices
✓	(g)	Choice and setting of protective and monitoring devices for protection against indirect contact and/or overcurrent
✗	(h)	Selection of equipment and protective measures appropriate to external influences
✓	(i)	Selection of appropriate functional switching devices

Inspected by *W. White* .... Date *18/4/2002* ....

Notes:

✓ to indicate an inspection has been carried out and the result is satisfactory
✗ to indicate an inspection has been carried out and the result was unsatisfactory
N/A to indicate the inspection is not applicable

SELV An extra-low voltage system which is electrically separate from earth and from other systems. The particular requirements of the Regulations must be checked (see Regulations 411-02 and 471-02)

Method of protection against direct contact - will include measurement of distances where appropriate

Obstacles - only adopted in special circumstances (see Regulations 412-04 and 471-06)

Placing out of reach - only adopted in special circumstances (see Regulations 412-05 and 471-07)

5. Use of Class II equipment - infrequently adopted and only when the installation is to be supervised (see Regulations 413-03 and 471-09)

6. Non-conducting locations - not applicable in domestic premises and requiring special precautions (see Regulations 413-04 and 471-10)

7. Earth-free local equipotential bonding - not applicable in domestic premises, only used in special circumstances (see Regulations 413-05 and 471-14)

8. Electrical separation (see Regulations 413-06 and 471-12)

Page *3* **of** *4*

Form 4
SCHEDULE OF TEST RESULTS

Form No *126/4*

Contractor: ...*County Electrics*...
Test Date: ...*18/4/2002*...
Signature ...*W White*...
Method of protection against indirect contact: ...
Equipment vulnerable to testing: ...

Address/Location of distribution board:
...*111 Any Street*...
...

Instruments
loop impedance: ...*LM.10*...
continuity: ...*LM.11*...
insulation: ...*LM.14*...
RCD tester: ...*LM.16*...

* Type of Supply: TN-S/TN-C-S/TT
* Ze at origin: ...... ohms
* PFC: *dis. board 0.24* kA
  at dis. board 1.0 ...

Description of Work: *House, Garage and Shed*

Circuit Description	Overcurrent Device *Short-circuit capacity: 6:kA*		Wiring Conductors			Continuity			Insulation Resistance		Polarity	Earth Loop Imped-ance Z_s	Functional Testing			Remarks
	type	Rating I_n	live	cpc		$R_1 + R_2$	R_2	Ring	Live/Live	Live/Earth			RCD time	Other	
		A	mm²	mm²		Ω	Ω		MΩ	MΩ		Ω	ms		
1	2	3	4	5		*6	*7	*8	*9	*10	*11	*12	*13	*14	15
Lights up	*fuse*	5	1.5	1.0	✓			10	8	✓	1.3	—	✓	*faulty pendants*
Lights down	"	5	1.5	1.0	✓	—	—	—	7	✓	1.2	—	✓	*faulty pendants*
Ring up	"	30	2.5	1.5	✓	—	✓	15	15	✓	0.7	—	✓	*faulty pendants*
Ring down	"	30	2.5	1.5	✓	—	✓	12	12	✓	0.8	—	✓	*no RCD*
Cooker	"	30	6.0	2.5	✓	—		20	15	✓	0.4	—	✓	
15A 10	"	15	2.5	1.5	✓	—		20	15	✓	0.4	—	✓	
Shower	"	30	6.0	2.5	✓	—		20	15	✓	0.3	—	✓	

Deviations from Wiring Regulations and special notes: *Main bonding cables too small,*

no supplementary bonding to bathroom,
no RCD to ground floor sockets
lighting pendants need replacing.

* See notes on schedule of test results

Page 4 of 4

NOTES ON SCHEDULE OF TEST RESULTS

* **Type of supply** is ascertained from the supply company or by inspection.

* **Z_e at origin.** When the maximum value declared by the electricity supplier is used, the effectiveness of the earth must be confirmed by a test. If measured the main bonding will need to be disconnected for the duration of the test.

* **Short-circuit capacity** of the device is noted, see Table 7.2A of the On-Site Guide or 2.7.15 of GN3

* **Prospective fault current (PFC).** The value recorded is the greater of either the short-circuit current or the earth fault current. Preferably determined by enquiry of the supplier.

The following tests, where relevant, shall be carried out in the following sequence:

Continuity of protective conductors, including main and supplementary bonding
Every protective conductor, including main and supplementary bonding conductors, should be tested to verify that they are continuous and correctly connected.

*6 **Continuity**
Where Test Method 1 is used, enter the measured resistance of the phase conductor plus the circuit protective conductor (R_1+ R_2).
See 10.3.1 of the On-Site Guide or 2.7.5 of GN3.
During the continuity testing (Test Method 1) the following polarity checks are to be carried out:
(a)　　every fuse and single-pole control and protective device is connected in the phase conductor only
(b)　　centre-contact bayonet and Edison screw lampholders have outer contact connected to the neutral conductor
(c)　　wiring is correctly connected to socket-outlets and similar accessories.
Compliance is to be indicated by a tick in polarity column 11.

*7　Where Test Method 2 is used, the maximum value of R_2 is recorded in column 7.
Where the alternative method of Regulation 413-02-12 is used for shock protection, the resistance of the circuit protective conductor R_2 is measured and recorded in column 7.
See 10.3.1 of the On-Site Guide or 2.7.5 of GN3.

*8 **Continuity of ring final conductors**
A test shall be made to verify the continuity of each conductor including the protective conductor of every ring final circuit.
See 10.3.2 of the On-Site Guide or 2.7.6 of GN3.

*9, *10 **Insulation Resistance**
All voltage sensitive devices to be disconnected or test between live conductors (phase and neutral) connected together and earth.
The insulation resistance between live conductors is to be inserted in column 9.
The minimum insulation resistance values are given in Table 10.1 of the On-Site Guide or Table 2.2 of GN3.
See 10.3.3(iv) of the On-Site Guide or 2.7.7 of GN3.

All the preceding tests should be carried out before the installation is energised.

*11 **Polarity**
A satisfactory polarity test may be indicated by a tick in column 11.
Only in a Schedule of Test Results associated with a Periodic Inspection Report is it acceptable to record incorrect polarity.

*12 **Earth fault loop impedance Z_s**
This may be determined either by direct measurement at the furthest point of a live circuit or by adding (R_1 + R_2) of column 6 to Z_e. Z_e is determined by measurement at the origin of the installation or preferably the value declared by the supply company used. $Z_s = Z_e + (R_1 + R_2)$. Z_s should be less than the values given in Appendix 2 of the On-Site Guide or App 2 of GN3.

*13 **Functional testing**
The operation of RCDs (including RCBOs) shall be tested by simulating a fault condition, independent of any test facility in the device.
Record operating time in column 13. Effectiveness of the test button must be confirmed.
See Section 11 of the On-Site Guide or 2.7.16 of GN3

*14　All switchgear and controlgear assemblies, drives, control and interlocks, etc must be operated to ensure that they are properly mounted, adjusted, and installed.
Satisfactory operation is indicated by a tick in column 14.

Earth electrode resistance
The earth electrode resistance of TT installations must be measured, and normally an RCD is required.
For reliability in service the resistance of any earth electrode should be below 200 Ω. Record the value on Forms 1, 2 or 6 as appropriate. See 10.3.5 of the On-Site Guide or 2.7.13 of GN3.

APPENDIX 8

STANDARD CIRCUIT ARRANGEMENT FOR HOUSEHOLD AND SIMILAR INSTALLATIONS

Introduction

This appendix gives advice on standard circuit arrangements for households and similar premises. The circuits satisfy the requirements of Chapter 43 for overload protection and Chapter 46 for isolation and switching, together with the requirements as regards current-carrying capacities of conductors prescribed in Chapter 52 for the selection and erection of wiring systems of BS 7671.

It is the responsibility of the designer and installer when adopting these circuit arrangements to take the appropriate measures to comply with the requirements of other chapters or sections which are relevant, such as Chapter 41, Protection against electric shock, Section 434, Protection against fault current, Chapter 54, Earthing arrangements and protective conductors, and the requirements of Chapter 52, Selection and erection of wiring systems, other than those concerning current-carrying capacities.

Circuit arrangements other than those detailed in this appendix are not precluded when specified by a suitably qualified electrical engineer, in accordance with the general requirements of Regulation 314-01-03.

The standard circuit arrangements are:

- Final circuits using socket-outlets complying with BS 1363-2 and fused connection units complying with BS 1363-4

- Cooker final circuits.

- Final radial circuits using socket-outlets complying with BS 4343 (BS EN 60309-2)

Final circuits using socket-outlets complying with BS 1363-2 and fused connection units complying with BS 1363-4

General

A ring or radial circuit, with spurs if any, feeds permanently connected equipment and an unlimited number of socket-outlets and fused connection units.

The floor area served by the circuit is determined by the known or estimated load and does not exceed the value given in Table 8A.

A single 30 A or 32 A ring circuit may serve a floor area of up to 433-02-04 100 m². Sockets for washing machines, tumble dryers and dishwashers should be located so as to provide reasonable sharing of the load in each leg of the ring, or consideration should be given to a separate circuit.

The number of socket-outlets is such as to ensure compliance with Regulation 553-01-07, each socket-outlet of a twin or multiple socket-outlet being regarded as one socket-outlet.

Diversity between socket-outlets and permanently connected equipment has already been taken into account in Table 8A and no further diversity should be applied.

TABLE 8A
Final circuits using BS 1363 socket-outlets and connection units

Type of circuit		Overcurrent protective device	Minimum conductor cross-sectional area*		Maximum floor area served
			Copper conductor thermoplastic or thermosetting insulated cables	Copper conductor mineral insulated cables	
		Rating A	mm²	mm²	m²
1	2	3	4	5	6
A1	Ring	30 or 32	2.5	1.5	100
A2	Radial	30 or 32	4	2.5	75
A3	Radial	20	2.5	1.5	50

*The tabulated values of conductor size may be reduced for fused spurs

Where two or more ring final circuits are installed the socket-outlets and permanently connected equipment to be served are to be reasonably distributed among the circuits.

Circuit protection

Table 8A is applicable for circuits protected by:

- fuses to BS 3036, BS 1361 and BS 88 and
- circuit-breakers

    Types B and C to BS EN 60898 or BS EN 61009-1 and

    BS EN 6 0947-2 and

    Types 1, 2 and 3 to BS 3871.

Conductor size

The minimum size of conductor cross-sectional area in the circuit and in non-fused spurs is given in Table 8A. However, the actual size of cable is determined by the current-carrying capacity for the particular method of installation, after applying appropriate correction factors from Appendix 6. The current-carrying capacity so calculated shall be not less than:

    20 A for circuit A1,

    30 A or 32 A for circuit A2 (i.e. the rating of the overcurrent protective device),

    20 A for circuit A3 (i.e. the rating of the overcurrent protective device).

The conductor size for a fused spur is determined from the total current demand served by that spur, which is limited to a maximum of 13 A.

When a fused spur serves socket-outlets the minimum conductor size is:

    1.5 mm^2 for cables with thermosetting or thermoplastic insulated cables, copper conductors,

    1 mm^2 for mineral insulated cables, copper conductors.

The conductor size for circuits protected by BS 3036 fuses is determined by applying the 0.725 factor of Regulation 433-02-03; that is, the current-carrying capacity must be at least 27 A for circuits A1 and A3, and 41 A for circuit A2.

Spurs

The total number of fused spurs is unlimited but the number of non-fused spurs should not exceed the total number of socket-outlets and items of stationary equipment connected directly in the circuit.

A non-fused spur feeds only one single or one twin or multiple socket-outlet or one permanently connected equipment. Such a spur is connected to a circuit at the terminals of a socket-outlet or junction box or at the origin of the circuit in the distribution board.

A fused spur is connected to the circuit through a fused connection unit, the rating of the fuse in the unit not exceeding that of the cable forming the spur and, in any event, not exceeding 13 A.

Permanently connected equipment

Permanently connected equipment is locally protected by a fuse complying with BS 1362 of rating not exceeding 13 A or by a circuit-breaker of rating not exceeding 16 A and of a type listed above and is controlled by a switch meeting the requirements of Regulation 476-03-04. A separate switch is not required where compliance with 476-03-04 is provided by the circuit-breaker.

Final radial circuits using 16 A socket-outlets complying with BS 4343 (BS EN 60309-2)

General

Where a radial circuit feeds equipment the maximum demand of which, having allowed for diversity, is known or estimated not to exceed the rating of the overcurrent protective device and in any event does not exceed 20 A, the number of socket-outlets is unlimited.

Circuit protection

The overcurrent protective device is to have a rating not exceeding 20 A.

Conductor size

The minimum size of conductor in the circuit is given in Table 8A. Where cables are grouped together the limitations of Para 7.2.1 and Appendix 6 apply.

Types of socket-outlets

Socket-outlets should have a rated current of 16 A and be of the type appropriate to the number of phases, circuit voltage and earthing arrangements. Socket-outlets incorporating pilot contacts are not included.

Cooker circuits in household or similar premises

The circuit supplies a control switch or a cooker unit complying with BS 4177, which may incorporate a socket-outlet.

The rating of the circuit is determined by the assessment of the current demand of the cooking appliance(s), and cooker control unit socket-outlet if any, in accordance with Table 1A of Appendix 1. A 30 or 32 A circuit is usually appropriate for household or similar cookers of rating up to 15 kW.

A circuit of rating exceeding 15 A but not exceeding 50 A may supply two or more cooking appliances where these are installed in one room. The control switch or cooker control unit should be placed within two metres of the appliance, but not directly above it. Where two stationary cooking appliances are installed in one room, one switch may be used to control both appliances provided that neither appliance is more than two metres from the switch. Attention is drawn to the need to provide discriminative operation of protective gear as stated in Regulation 533-01-06.

Water and space heating

Water heaters fitted to storage vessels in excess of 15 litres capacity, or permanently connected heating appliances forming part of a comprehensive space heating installation, are to be supplied by their own separate circuit.

Immersion heaters are to be supplied through a switched cord-outlet-connection-unit complying with BS 1363-4.

Heights of switches and sockets

The Building Regulations require switches and socket-outlets in dwellings to be installed so that all persons including those whose reach is limited can easily use them. A way of satisfying the requirement is to install switches and socket-outlets in habitable rooms at a height of between 450 mm and 1200 mm 553-01-06 from the finished floor level - see Figure 8A. Unless the dwelling is for persons whose reach is limited the requirements would not apply to kitchens and garages but specifically only to rooms that visitors would normally use.

The Building Regulations are not applicable in Scotland where the Building Standards (Scotland) Regulations apply. The Scottish regulations do not have specific minimum heights for accessories, installations are required to generally comply with BS 7671.

Fig 8A: Height of switches, sockets etc

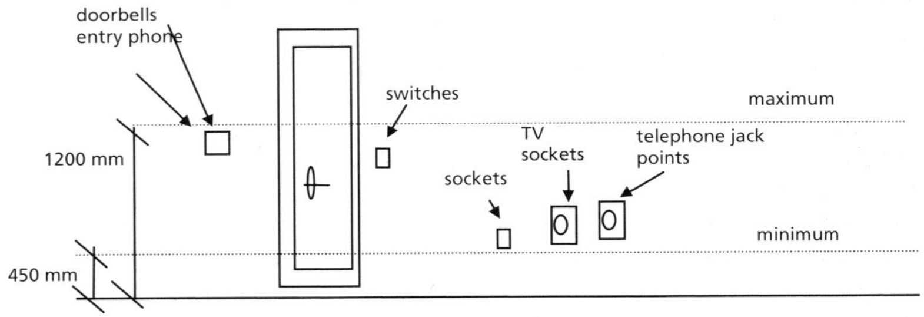

From Approved Document M, 1999 edition Section 8.

Number of socket-outlets

Sufficient socket-outlets are required to be installed so that all equipment likely to be used can be supplied from a reasonably accessible socket-outlet, taking account of the length of flexible cords normally fitted to appliances and luminaires. (Regulation 553-01-07). Table 8B provides guidance on the number of 553-01-07 socket-outlets that are likely to meet this requirement.

TABLE 8B
Recommended provision of socket-outlets
(All socket-outlets are twin)

Location	No. of outlets	Notes
Lounge	6 to 10	(1) (2) (3) (9)
Dining	3	
Kitchen	6 to 10	(3) (4) (5) (9)
Double Bedroom	4 to 6	(3)
Single Bedroom	4 to 6	(3) (6) (9)
Bedsitter	4	
Hall	2	(7)
Stairs/Landing	1	
Loft	1	(7)
Study/Home office	6	(7) (8) (9)
Garage	2	
Utility Room	2	(5)

(This table was prepared with the kind assistance of the ECA, Select, NHBC, CDA and EIEMA).

Notes:

(1) The number of outlets depends on the size of the room.

(2) Two twin socket-outlets should be located close to the TV aerial outlet to allow for TV, video etc. and ancillary equipment supplies

(3) Larger dwellings will require proportionally more socket-outlets than smaller dwellings.

(4) Kitchens should be fitted with socket-outlets above work surfaces as well as specific socket-outlets for built in appliances.

(5) A lower number of socket-outlets may be appropriate in a kitchen where the washing machine, dryer, freezer etc. are expected to be installed in a separate utility room.

(7) One twin socket-outlet should be installed near any telephone outlet to supply mains powered telecommunication equipment.

(8) The provisions for an office at home may require more consideration with the user to identify and locate all necessary equipment than is the case with an ordinary domestic installation.

(9) The use of IT and other electrical equipment with high earth protective conductor currents may require the application of Regulation 607 to accommodate cumulative leakage currents.   607-02-06

APPENDIX 9

RESISTANCE OF COPPER AND ALUMINIUM CONDUCTORS

To check compliance with Regulation 434-03-03 and/or 434-03-03 Regulation 543-01-03, i.e. to evaluate the equation $s^2 = I^2 t/k^2$, it 543-01-03 is necessary to establish the impedances of the circuit conductors to determine the fault current I and hence the protective device disconnection time t.

Fault current $I = U_o/Z_s$

where

U_o is the nominal voltage to earth,
Z_s is the earth fault loop impedance.

$Z_s = Z_e + R_1 + R_2$

where

Z_e is that part of the earth fault loop impedance external to the circuit concerned,

R_1 is the resistance of the phase conductor from the origin of the circuit to the point of utilization,

R_2 is the resistance of the protective conductor from the origin of the circuit to the point of utilization.

Similarly, in order to design circuits for compliance with BS 7671 limiting values of earth fault loop impedance to those given in Tables 41B1, 41B2 and 41D of BS 7671, or for compliance with the limiting values of the circuit protective conductor given in Table 41C, it is necessary to establish the relevant impedances of the circuit conductors concerned at their operating temperature.

Table 9A gives values of (R1 + R2) per metre for various combinations of conductors up to and including 50 mm^2 cross-sectional area. It also gives values of resistance (milliohms) per metre for each size of conductor. These values are at 20 °C.

TABLE 9A
Value of resistance/metre for copper and aluminium conductors and of R₁ + R₂ per metre at 20 °C in milliohms/metre

Cross-sectional area (mm²)		Resistance/metre or (R1 + R2)/metre (mΩ/m)	
Phase conductor	Protective conductor	Copper	Aluminium
1	—	18.10	
1	1	36.20	
1.5	—	12.10	
1.5	1	30.20	
1.5	1.5	24.20	
2.5	—	7.41	
2.5	1	25.51	
2.5	1.5	19.51	
2.5	2.5	14.82	
4	—	4.61	
4	1.5	16.71	
4	2.5	12.02	
4	4	9.22	
6	—	3.08	
6	2.5	10.49	
6	4	7.69	
6	6	6.16	
10	—	1.83	
10	4	6.44	
10	6	4.91	
10	10	3.66	
16	—	1.15	1.91
16	6	4.23	—
16	10	2.98	—
16	16	2.30	3.82
25	—	0.727	1.20
25	10	2.557	—
25	16	1.877	—
25	25	1.454	2.40
35	—	0.524	0.87
35	16	1.674	2.78
35	25	1.251	2.07
35	35	1.048	1.74
50	—	0.387	0.64
50	25	1.114	1.84
50	35	0.911	1.51
50	50	0.774	1.28

TABLE 9B
Ambient temperature multipliers to Table 9A

Expected ambient temperature	Correction factor note
5 °C	0.94
10 °C	0.96
15 °C	0.98
20 °C	1.00
25 °C	1.02

Note:
The correction factor is given by:
{1 + 0.004 (ambient temp - 20 °C}
where 0.004 is the simplified resistance coefficient per °C at 20 °C given by BS 6360 for copper and aluminium conductors.

For verification purposes the designer will need to give the values of the phase and circuit protective conductor resistances at the ambient temperature expected during the tests. This may be different from the reference temperature of 20 °C used for Table 9A. The correction factors in Table 9B may be applied to the Table 9A values to take account of the ambient temperature (for test purposes only).

Multipliers for conductor operating temperature

Table 9C gives the multipliers to be applied to the values given in Table 9A for the purpose of calculating the resistance at maximum operating temperature of the phase conductors and/or circuit protective conductors in order to determine compliance with, as applicable:

(a)   earth fault loop impedance of Table 41B1, Table 41B2 or Table 41D of BS 7671

Table 41B1
Table 41B2
Table 41D

(b)   earth fault loop impedance and resistance of protective conductor of Table 41C of BS 7671.

Table 41C

Where it is known that the actual operating temperature under normal load is less than the maximum permissible value for the type of cable insulation concerned (as given in the Tables of current-carrying capacity) the multipliers given in Table 9C may be reduced accordingly.

TABLE 9C Standard devices

Multipliers to be applied to Table 9A to calculate conductor resistance at maximum operating temperature

Table 41B1
Table 41B2
Table 41C
Table 41D

Conductor Installation	Conductor Insulation		
	70 °C thermoplastic (pvc)	85 °C thermosetting (rubber)	90 °C thermosetting
Not incorporated in a cable and not bunched - note 1	1.04	1.04	1.04
Incorporated in a cable or bunched - note 2	1.20	1.26	1.28

Table 54B

Table 54C

Note 1 See Table 54B of BS 7671: applies where the protective conductor is not incorporated or bunched with cables, or for bare protective conductors in contact with cable covering.

Table 54B

Note 2 See Table 54C of BS 7671: applies where the protective conductor is a core in a cable or is bunched with cables.

Table 54C

The multipliers given in Table 9C for both copper and aluminium conductors are based on a simplification of the formula given in BS 6360, namely that the resistance-temperature coefficient is 0.004 per deg C at 20 °C.

APPENDIX 10

PROTECTIVE CONDUCTOR SIZING

TABLE 10A
Main earthing and main equipotential bonding conductor sizes (copper equivalent) for TN-S and TN-C-S supplies

Phase conductor or neutral conductor of PME supplies	mm²	4	6	10	16	25	35	50	70	
Earthing conductor not buried or buried protected against corrosion and mechanical damage see notes	mm²	6	6	10	16	16	16	25	35	542-03-01 543-01-01
Main equipotential bonding conductor see notes	mm²	6	6	6	10	10	10	16	25	547-02-01
Main equipotential bonding conductor for PME supplies (TN-C-S)	mm²	10	10	10	10	10	10	16	25	Table 54H

Notes to Table 10A:

1. Protective conductors (including earthing and bonding conductors) of 10 mm² cross-sectional area or less shall be copper.   543-02-03

2. Regional electricity companies may require a minimum size of earthing conductor at the origin of the supply of 16 mm² copper or greater for TN-S and TN-C-S supplies

3. Buried earthing conductors must be at least :   542-03-01   Table 54A

   25 mm² copper if not protected against mechanical damage or corrosion
   50 mm² steel if not protected against mechanical damage or corrosion
   16 mm² copper if not protected against mechanical damage but protected against corrosion
   16 mm² coated steel if not protected against mechanical damage but protected against corrosion

4. The electricity supplier should be consulted when in doubt.

TABLE 10B
Supplementary bonding conductors

547-03

Size of circuit protective conductor mm²	Minimum cross-sectional area of supplementary bonding					
	Exposed-conductive-part to extraneous-conductive-part		Exposed-conductive-part to exposed-conductive-part		Extraneous-conductive-part to extraneous-conductive-part (1)	
	mechanically protected mm²	not mechanically protected mm²	mechanically protected mm²	not mechanically protected mm²	mechanically protected mm²	not mechanically protected mm²
	1	2	3	4	5	6
1.0	1.0	4.0	1.0	4.0	2.5	4.0
1.5	1.0	4.0	1.5	4.0	2.5	4.0
2.5	1.5	4.0	2.5	4.0	2.5	4.0
4.0	2.5	4.0	4.0	4.0	2.5	4.0
6.0	4.0	4.0	6.0	6.0	2.5	4.0
10.0	6.0	6.0	10.0	10.0	2.5	4.0
16.0	10.0	10.0	16.0	16.0	2.5	4.0

Note 1: If one of the extraneous-conductive-parts is connected to an exposed-conductive-part, the bond must be no smaller than that required for bonds between exposed-conductive-parts - columns 3 or 4.

TABLE 10C

Copper earthing conductor cross-sectional area (csa) for TT supplies for earth fault loop impedances not less than 1 ohm (Ω)

Buried			Not buried		
Unprotected	Protected against corrosion	Protected against corrosion and mechanical damage	Unprotected	Protected against corrosion	Protected against corrosion and mechanical damage
mm²	mm²	mm²	mm²	mm²	mm²
25	16	2.5	4	4	2.5

Notes:

1. Protected against corrosion by a sheath.

2. For impedances less than 1 ohm determine as per Regulation 543-01-02.

3. The main equipotential bonding conductor shall have a cross-sectional area of not less than half that required for the earthing conductor and not less than 6 mm².

THIS PAGE HAS BEEN LEFT BLANK INTENTIONALLY

PICTORIAL INDEX

A four part pictorial index follows comprising the following schematic diagrams:

To use the index, turn to the relevant index, and find paragraph references against the appropriate schematic drawing.

Index (i) THE INSTALLATION

basic information 1.2

information
structure
wiring
bonding

isolation and
switching Sec 5
functional
switching 5.2
isolation 5.1(iii)
emergency
switch 5.4

labelling isolation
by more than one
device 6.1(viii)

labelling 6.1(i)
and (ii) unexpected
voltage exceeding
230v

labelling 6.1(iii)
presence of
different nominal
voltages
DATA cables
7.3.3

fireman's
switch 5.4

neon sign

external equipment
IP code, 8.3(iv)

band I
circuit 9.2.2(ix)

particular
attention
is required
8.3

conventional
final circuits
Table 7.1
assumed
conditions
7.1

E M M A

garage
or shed

choice of
protective devices
7.2.5

bathrooms and
showers 8.1

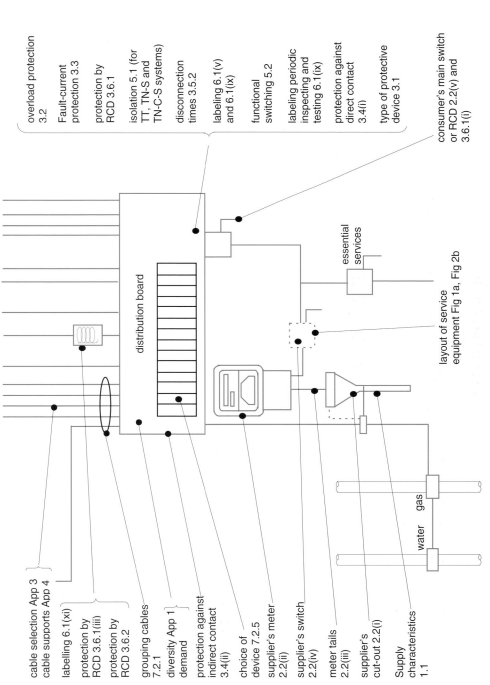

overload protection 3.2

Fault-current protection 3.3

protection by RCD 3.6.1

isolation 5.1 (for TT, TN-S and TN-C-S systems)

disconnection times 3.5.2

labeling 6.1(v) and 6.1(ix)

functional switching 5.2

labeling periodic inspecting and testing 6.1(ix)

protection against direct contact 3.4(i)

type of protective device 3.1

consumer's main switch or RCD 2.2(v) and 3.6.1(i)

essential services

layout of service equipment Fig 1a, Fig 2b

distribution board

water gas

cable selection App 3
cable supports App 4

labelling 6.1(xi)

protection by RCD 3.6.1(iii)

protection by RCD 3.6.2

grouping cables 7.2.1

diversity App 1
demand

protection against indirect contact 3.4(ii)

choice of device 7.2.5

supplier's meter 2.2(ii)

supplier's switch 2.2(iv)

meter tails 2.2(iii)

supplier's cut-out 2.2(i)

Supply characteristics 1.1

167

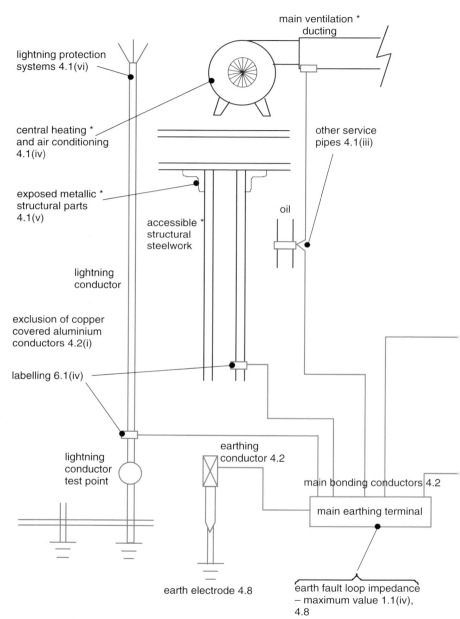

lightning protection
systems 4.1(vi)

main ventilation *
ducting

central heating *
and air conditioning
4.1(iv)

other service
pipes 4.1(iii)

exposed metallic *
structural parts
4.1(v)

accessible *
structural
steelwork

oil

lightning
conductor

exclusion of copper
covered aluminium
conductors 4.2(i)

labelling 6.1(iv)

lightning
conductor
test point

earthing
conductor 4.2

main bonding conductors 4.2

main earthing terminal

earth electrode 4.8

earth fault loop impedance
– maximum value 1.1(iv),
4.8

* metal parts only require main bonding when they
  are extraneous-conductive-parts, see Section 4

Index (ii) **BONDING AND EARTHING**

See also Fig 4a, Fig 4b and Fig 4c

information
structure
wiring
bonding

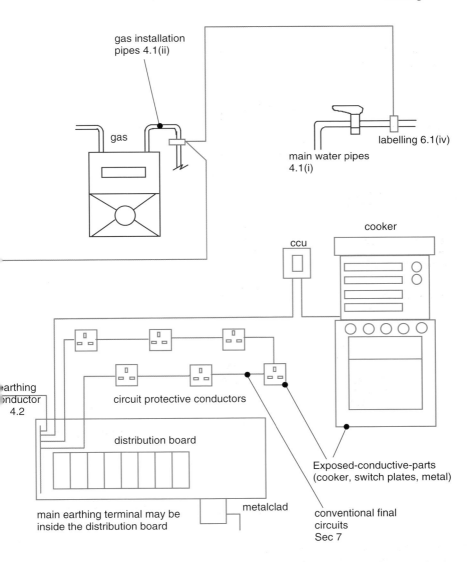

gas installation
pipes 4.1(ii)

gas

labelling 6.1(iv)

main water pipes
4.1(i)

cooker

ccu

earthing
conductor
4.2

circuit protective conductors

distribution board

Exposed-conductive-parts
(cooker, switch plates, metal)

main earthing terminal may be
inside the distribution board

metalclad

conventional final
circuits
Sec 7

Index (iii) SPECIAL LOCATIONS AND RCDs

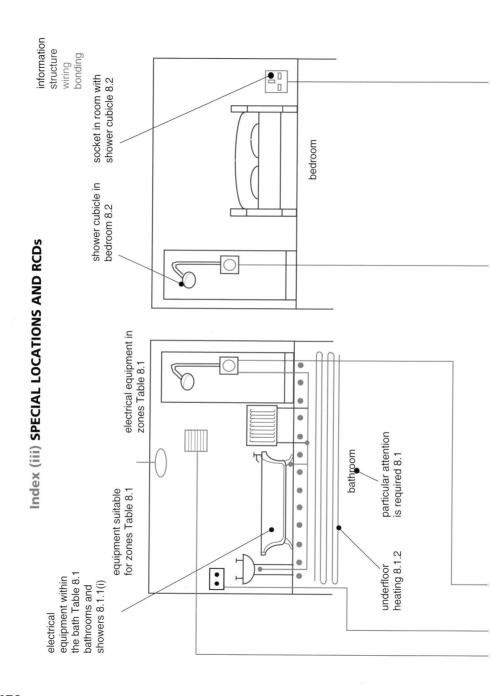

information
structure
wiring
bonding

socket in room with
shower cubicle 8.2

shower cubicle in
bedroom 8.2

bedroom

electrical equipment in
zones Table 8.1

equipment suitable
for zones Table 8.1

electrical
equipment within
the bath Table 8.1
bathrooms and
showers 8.1.1(i)

bathroom

particular attention
is required 8.1

underfloor
heating 8.1.2

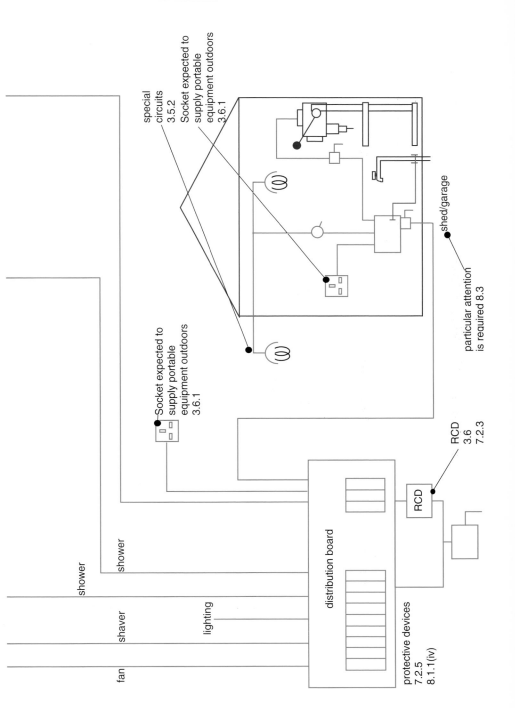

fan

shaver

shower

shower

lighting

special
circuits
3.5.2
Socket expected to
supply portable
equipment outdoors
3.6.1

Socket expected to
supply portable
equipment outdoors
3.6.1

shed/garage

particular attention
is required 8.3

distribution board

protective devices
7.2.5
8.1.1(iv)

RCD

RCD
3.6
7.2.3

Index (iv) INSPECTION AND TESTING

information
structure
wiring
bonding

continuity
testing 10.3.1
and 10.3.2

polarity testing
10.3.4

insulation
resistance
testing 10.3.3

loop impedance
App 2

distribution board

Initial testing
guidance notes
Sec 10

safety during
testing 10.1

sequence of
tests 10.2

test procedures
10.3

test checklist
9.3.1

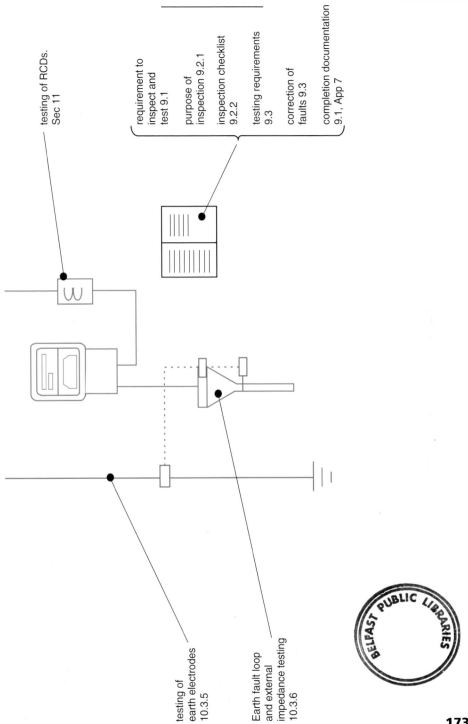

testing of RCDs.
Sec 11

requirement to
inspect and
test 9.1

purpose of
inspection 9.2.1

inspection checklist
9.2.2

testing requirements
9.3

correction of
faults 9.3

completion documentation
9.1, App 7

testing of
earth electrodes
10.3.5

Earth fault loop
and external
impedance testing
10.3.6

Index (v) **ALPHABETICAL**

See page 165 for pictorial index